3

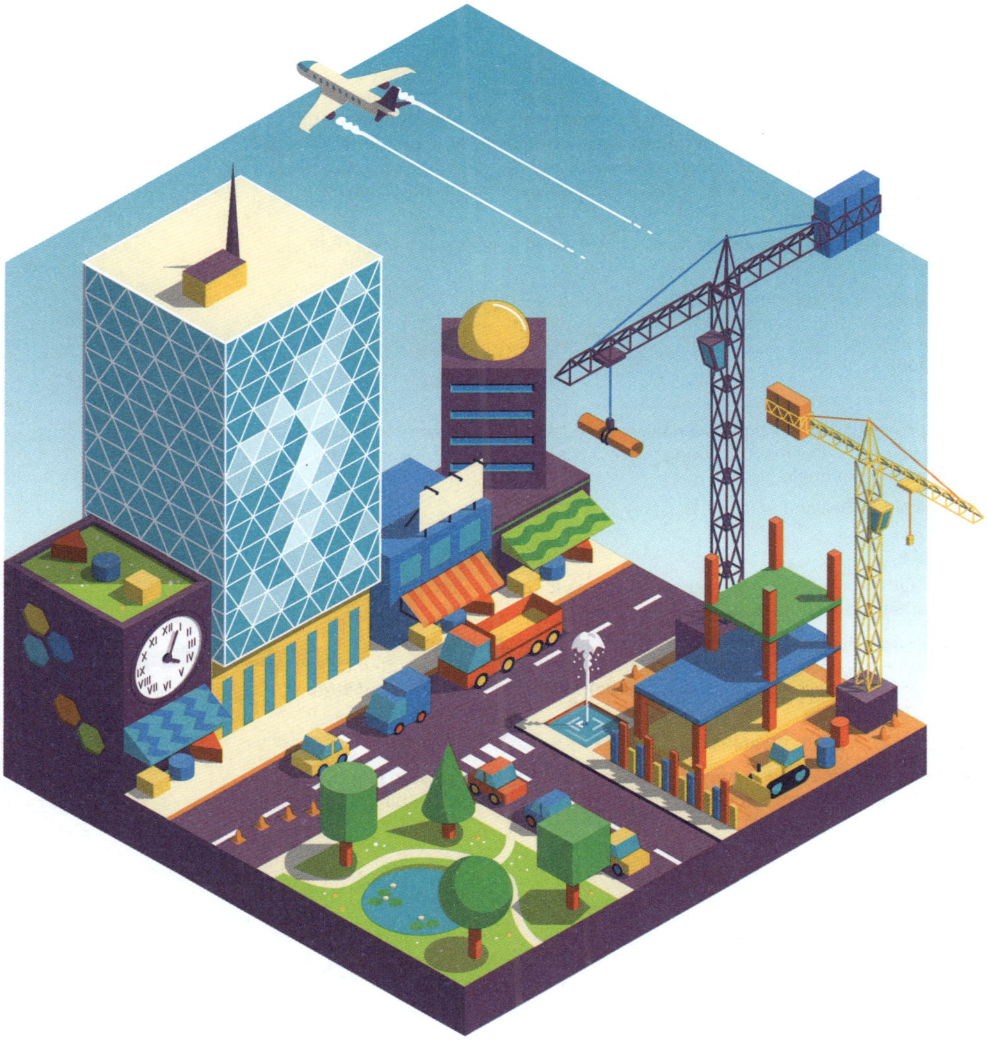

Pupil Book

Karen Morrison
Lisa Greenstein

OXFORD
UNIVERSITY PRESS

OXFORD
UNIVERSITY PRESS

Great Clarendon Street, Oxford, OX2 6DP, United Kingdom

Oxford University Press is a department of the University of Oxford.

It furthers the University's objective of excellence in research, scholarship, and education by publishing worldwide. Oxford is a registered trade mark of Oxford University Press in the UK and in certain other countries.

British Library Cataloguing in Publication Data

Data available

ISBN: 978-1-382-01002-3

1 3 5 7 9 10 8 6 4 2

Paper used in the production of this book is a natural, recyclable product made from wood grown in sustainable forests. The manufacturing process conforms to the environmental regulations of the country of origin.

Printed in Great Britain by Bell and Bain Ltd, Glasgow

Acknowledgements

The publisher and authors would like to thank the following for permission to use photographs and other copyright material:

Cover: Matthieu Nivesse. Photos: **p5(a):** Silver Wings SS/Shutterstock; **p5(b):** Anna_Pustynnikova/Shutterstock; **p5(c):** New Africa/Shutterstock; **p5(d):** KETriKET/Shutterstock; **p33:** Samira Mian; **p52:** Steve Allen/Shutterstock; **p58(bg):** Dudaeva/Shutterstock; **p58(fg):** Ermak Oksana/Shutterstock.

Artwork by Aviel Basil, Q2A Media, Pantek Media, and OKS Prepress.

Every effort has been made to contact copyright holders of material reproduced in this book. Any omissions will be rectified in subsequent printings if notice is given to the publisher.

Contents

Think maths

Seeing in different ways

 Think and share

What numbers do you see? What shapes do you see?

What is similar? What is different?

1 **Estimate** the number of holes in the spoon. Do not count! Are there about 6, about 60 or about 600 holes? How did you estimate?

2 Work in pairs. Find three different ways to work out the number of holes in the spoon. Use:

- only counting

- counting and adding

- counting and multiplying.

Share your ideas with the class. Which way was the easiest?

3 Look at the pancakes.

a Why is it impossible to count the pancakes?

b Explain why both these sentences are true:

- There is at least one pancake on the plate.

- There is more than one pancake on the plate.

c How could we rearrange the pancakes to make them easier to count?

4 **a** If you add two more plums to the bowl, what shape will they make?

b Draw dots in the same arrangement as the plums. Join the dots in different ways. What shapes or patterns can you make?

Maths mindset

- Maths is a language that helps us to describe and understand the world around us.

- We find out more by asking questions.

- There are many different ways to solve a problem.

- Maths problems take time. There is no need to rush.

- Sometimes, part of finding an answer is working out what the question means.

Sometimes learning maths seems hard. Sometimes we make mistakes. Sometimes we need to ask questions or work with a partner. Your maths mindset is all the things you believe about maths. It is OK to struggle, to make mistakes and to ask questions. This is part of maths.

1 Neither of these pupils knows the answer to the maths problem.

We don't know the answer. We can't do maths.

$5 + \square = 28$

We don't know ... yet. But we can work it out together. Let's keep trying.

a What do the pupils' statements tell you about their mindsets? Are their mindsets the same or different?

b Which mindset can help you in your maths work?

2 How do you feel when you make mistakes? Share ideas with a partner.

3 When you make a mistake in maths, what do you do?

I give up.

I ask questions to find out more.

I know that I have found one way that doesn't work, so I try a different way.

I ask someone to tell me the answer.

I work with a partner.

I try to work out how I went wrong.

4 As a class, discuss actions that are useful when you make mistakes in maths.

Our maths statements

Sometimes it feels good when maths problems are easy. But your brain grows the most when the problems are hard.

1 What would you say to each pupil?

This is too hard for me!

I don't want to share my answer in case I look stupid.

Just tell me whether I got the answer right or wrong.

I only want to do the easiest questions.

I don't know the answer, so I'm not going to try.

I feel silly when my answer is wrong.

2 This class made a poster with statements that are helpful and not helpful.

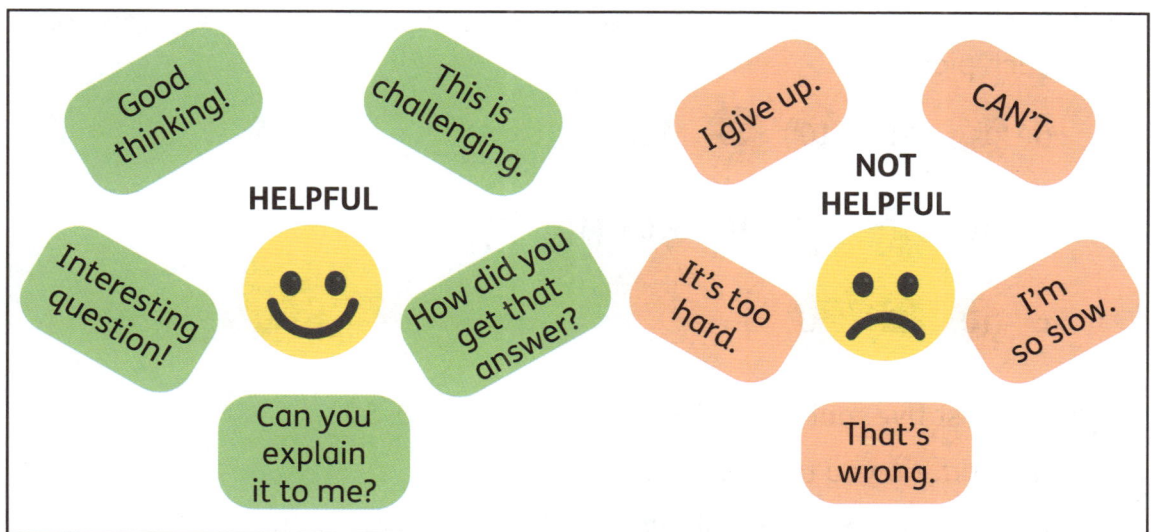

Write your own 'helpful' and 'not helpful' statements to help everyone in your class do well at maths. You could make a poster.

▶ *Workbook page 4*

Number and place value

Revise numbers to 100

> ### Think and share
>
> This ten frame has 10 holes. Some holes are covered with blue counters.
>
>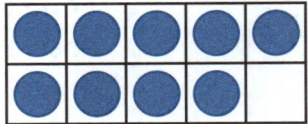
>
> - If each counter is worth 1, what number sentences can you make?
> - If each counter is worth 10, what number sentences can you make?
>
> A **number sentence** is a combination of numbers and signs.

1 Count on in tens from 0 to 100. Then count back in tens from 100 to 0.

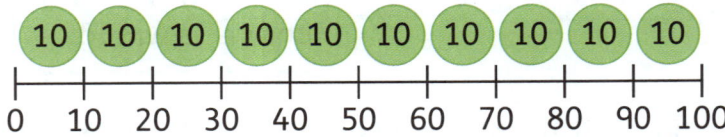

2 Each counter is worth 10. Write the number and the number name for each picture.

a (10) (10) (10) (10) (10)

b (10) (10) (10) (10) (10) (10) (10) (10)

c (10) (10) (10)

3 The first five multiples of 10 are 10, 20, 30, 40, 50.

a What do you think a **multiple** of 10 is?

b Write some more multiples of 10. What do all multiples of 10 have in common?

4 Adi says: 'Any multiple of 10 is also a multiple of 5 and a multiple of 2.' Discuss this with a partner. How do you think Adi knows this is true?

Understanding place value

We can use the digits 0, 1, 2, 3, 4, 5, 6, 7, 8 and 9 to write any number. The place of a digit in a number tells us the digit's value.

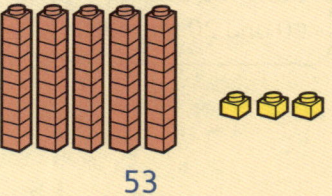

5<u>3</u>

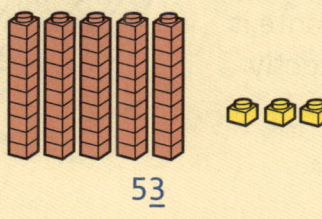

5<u>3</u>

Tens	Ones
5	3

The 5 is in the tens place.
It has a value of 50.

The 3 is in the ones place.
It has a value of 3.

1 Write the value of the underlined digit.

a

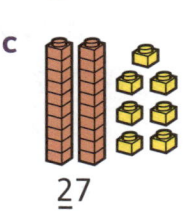

1<u>4</u>

b

35

c

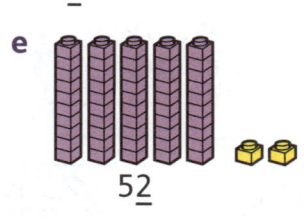

2<u>7</u>

d

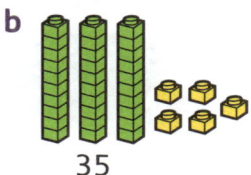

<u>8</u>9

e

5<u>2</u>

f

9<u>1</u>

2 Write these numbers using digits:

a six tens and seven ones

b eight ones and four tens

c nine tens and zero ones

d three ones and three tens

Problem solving

 How many different ways can you make 10?

3 I am a 2-digit number. My tens and ones digits add up to 10, and the **difference** between my digits is 4. What two numbers could I be?

Estimate and count

The estimate '10 to 20' is a **range**.
It tells you that there are between
10 and 20 marbles. This estimate is
correct, because there are exactly
15 marbles.

It is useful to estimate before
you count.

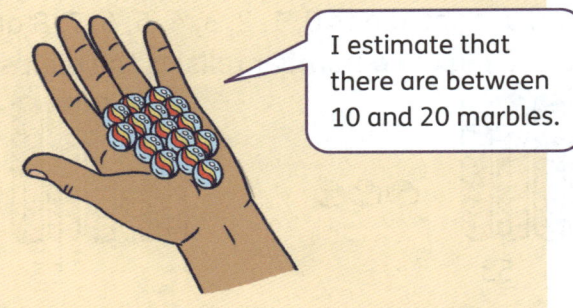

I estimate that
there are between
10 and 20 marbles.

1 Estimate and then count. Give your estimate as a range.

You will need:

- beans
- big buttons or counters
- small containers (a cup, a bowl, a spoon and a jar).

How many ...

a   beans in a handful?

b   buttons in a handful?

c   beans in a cup?

d   buttons on a spoon?

e   buttons in a bowl?

f   beans in a jar?

2 Estimate how many tomatoes are in each box. Give your estimate as a range.

Use smaller groups to help you estimate.

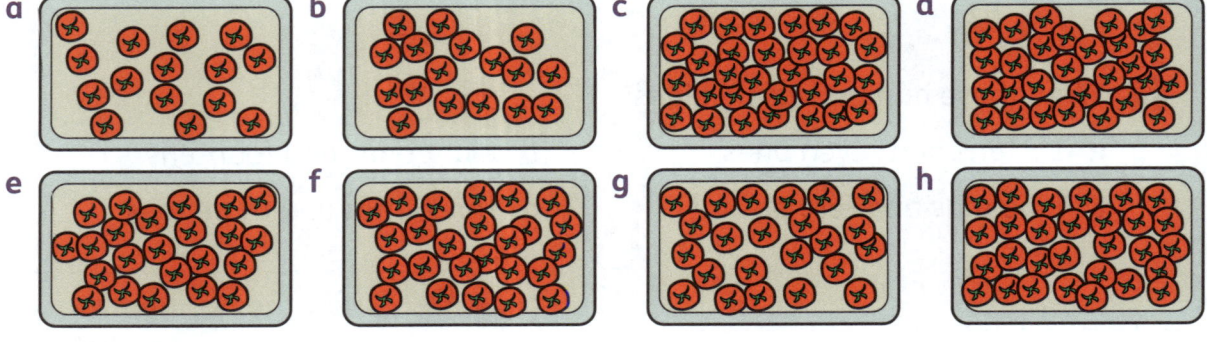

a   b   c   d

e   f   g   h

Tens and hundreds

10 tens are equal to 100.

(10) (10) (10) (10) (10) (10) (10) (10) (10) (10) = (100)

10 tens   = 1 hundred

(10) (10) (10) (10) (10)   (10) (10) (10) (10) | H | T | O |
(10) (10) (10) (10) (10)   +   = | 1 | 4 | 0 |

10 tens   +   4 tens   =   14 tens

This is a **place-value table**. It shows each place in the number. Can you see how many there are in each place?

Hundreds	Tens	Ones
1	4	0

1 Write the number and number name.

a (1 hundred) (2 tens) (1 one)   b (2 hundreds) (0 tens) (5 ones)

2 Write the number and the number name for each group of tens.

a 10 tens   b 12 tens   c 28 tens

d 49 tens   e 50 tens   f 99 tens

150 = 1 hundred + 5 tens + 0 ones. This is called expanding the number. We can also write it like this: 150 = 100 + 50 + 0

3 Expand these numbers. Write how many hundreds and tens there are.

a 470   b 340   c 560

Problem solving

$1 = 100c

4 a A shopkeeper has 830 pens. How many packs of 10 pens can she make?

b The shopkeeper sells stickers for 10c each. How many stickers must she sell to make $20?

➡ *Workbook page 5*

Hundreds, tens and ones

This picture shows 1 hundred, 2 tens and 3 ones.

The number is 123.

We say one hundred and twenty-three.

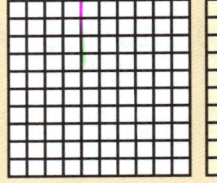

Hundreds	Tens	Ones
1	2	3

1 Draw a place-value table to show how many hundreds, tens and ones there are in each picture.

Then write and say the number.

a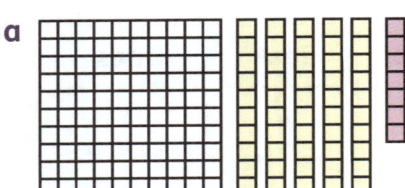

b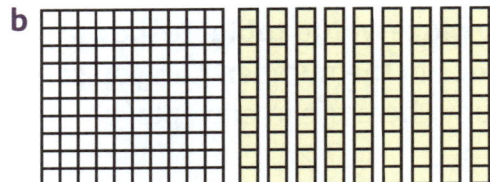

c

d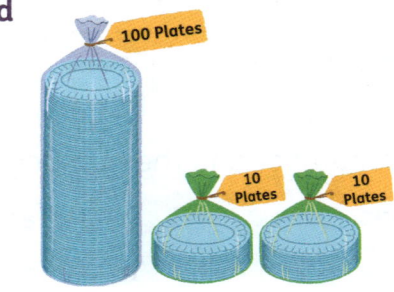

2 Write an addition and a total for each set of counters. The first one is done for you.

a

100 100 100

10 10

1 1 1 1

300 + 20 + 4 = 324

b

10

100 100 100

100 10 100 10

1 1 1

1 1 1

c

10 100 100

10 1 100

10 1

10

3 Write the answer to each addition.

a 200 + 30 + 9

b 50 + 7 + 800

c 1 + 700 + 40

➡️ *Workbook page 6 and page 7*

Make 3-digit numbers

2 hundreds (H), 4 tens (T) and 3 ones (O).

We can write 200 + 40 + 3

We can also make this number using place-value cards like this:

$$\boxed{2\,0\,0}\!/ \quad \boxed{4\,0}\!/ \quad \boxed{3}\!/ \;=\; \boxed{2\,4\,3}\!/$$

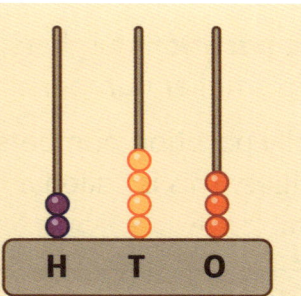

1 Use place-value cards to make each number. Write and say the number.

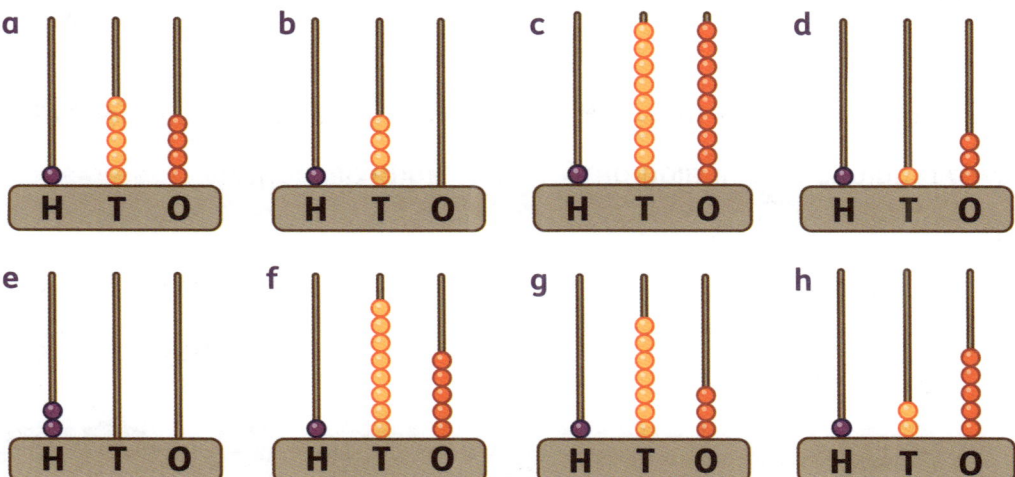

2 Write the value of the underlined red digit in each number.

a 1<u>2</u>8　　　　b <u>2</u>45　　　　c <u>3</u>03　　　　d 13<u>7</u>

e 4<u>9</u>0　　　　f 32<u>9</u>　　　　g 18<u>7</u>　　　　h <u>7</u>81

3 Work with a partner. What are the mistakes in each set of place-value cards?

Write and say each number correctly.

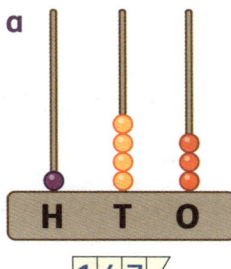
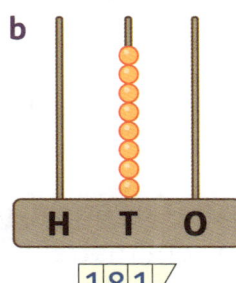
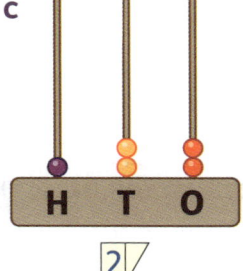
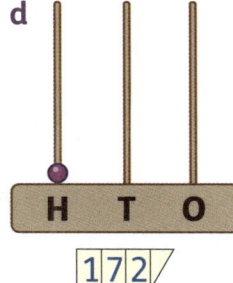

a $\boxed{1\,4\,7}\!/$　　b $\boxed{1\,8\,1}\!/$　　c $\boxed{2}\!/$　　d $\boxed{1\,7\,2}\!/$

➡ *Workbook page 8*

Partition numbers

When we **partition** numbers, we break them into smaller numbers.

We usually partition numbers into place values. This is called a **regular partition**.

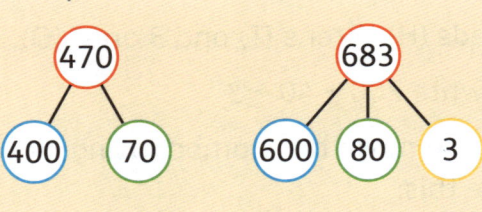

We can also partition numbers in different ways. This is called an **irregular partition**.

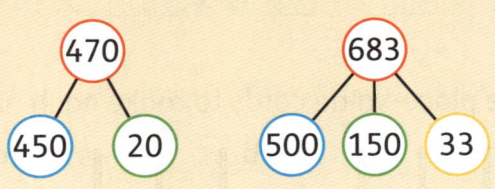

1 Draw diagrams to show three different ways to partition the number 375.

2 Draw a regular partition diagram for each group of counters.

a

100 100 10
10
100 100
1

b

10 100 1 1 1
10 100 1 1 1
10 1 1 1
10

c

1 1 10 100 100 100
1 1 10 10 100 100 100
1 10 10 10 100 100 100

d

100 10 100
10
10
10 100
100
100 10 10

Problem solving

1 m = 100 cm

3 Last year a tree was 3 m tall. Now the tree has grown another 25 cm. How tall is the tree now? Give your answer in centimetres.

➡ *Workbook page 9*

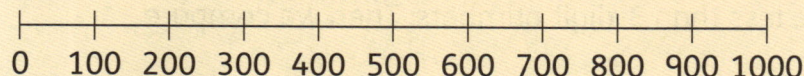

Number lines

We can use **number lines** to show the position of numbers.

```
├──┼──┼──┼──┼──┼──┼──┼──┼──┼──┤
0  100 200 300 400 500 600 700 800 900 1000
```

This number line counts in hundreds to 1000. 1000 is the number that comes after 999. We say one thousand. Can you see how many hundreds make 1000?

Number lines make it easy to compare numbers. We can see which numbers are less than or greater than other numbers.

- As you move left towards 0, the numbers get smaller in value.
  So 200 < 500

- As you move right towards 1000, the numbers get larger in value.
  So 800 > 300

1 Write the missing numbers for each number line.

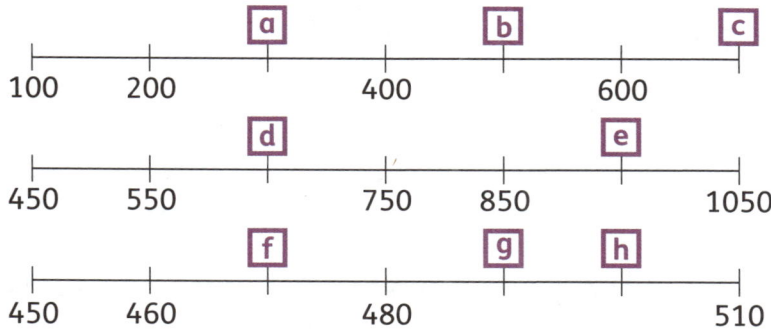

2 Look at this number line.

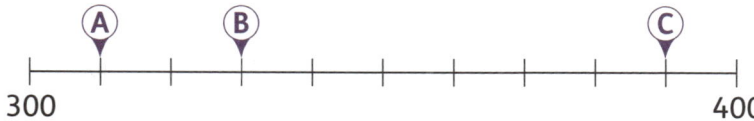

a Complete this sentence: This number line shows numbers between ___ and ___ .

b Where would you place 350 on this line? Show your partner and explain why.

c What numbers would you write at points **A**, **B** and **C**?

d Where would you place 395 on this line? Show your partner and explain why.

➡ Workbook page 10

Order numbers

When we **order** numbers, we first look at how many digits each number has. All 2-digit numbers are less than 3-digit numbers. Then we compare the digits in each place.

1 Order the numbers on each sticker. Write the numbers from smallest to greatest.

a
101 96 89
11   26 48
62 163

b
75 120 104
138 141 165
134 86

c
111 121 132
144 194 133
157 139

2 Order the numbers in each set. Write the numbers from greatest to smallest.

a 162 | 17 | 63 | 29 | 35 | 82 | 51 | 21

b 14 | 52 | 24 | 37 | 145 | 92 | 135 | 42

c 99 | 36 | 108 | 47 | 61 | 125 | 22 | 173

Problem solving

3 A teacher has given a group of pupils five number cards each. The pupils have put the cards in order from smallest to greatest.

Marie 497 □ 501 □ 504   **Josh** □ 345 347 □ 351

Sammy 400 □ 403 □ 410   **Nika** 698 □ 702 □ 710

Malika 800 810 □ □ 830   **Sheldon** 900 □ 908 □ 912

Two cards in each set are face down, so you cannot see the numbers.

Write the smallest number and the greatest number that could be on each face-down card.

Explain to your partner how you chose those numbers.

➡ *Workbook page 11*

Compare numbers

I have these three number cards.

The greatest number I can make is 652.

The smallest number I can make is 256.

I can compare the numbers using **<** or **>** signs.

652 > 256   256 < 652

1 Make the greatest number and the smallest number with each set of cards. Write the numbers using the **>** sign.

a 2 7 1   b 5 3 6   c 1 4 2

d 9 8 2   e 7 4 6   f 8 1 4

2 Look at your answers for question 1. Use the same digits to make a number that is between the greatest number and the smallest number.

3 Make the greatest number and the smallest number with each set of cards. Write the numbers using the **<** sign.

a 7 6 1   b 4 3 9   c 2 6 6

d 3 5 1   e 3 8 7   f 1 2 2

4 Look at your answers for question 3. Use the same digits to make a number that is between the smallest number and the greatest number.

Problem solving

You can use a number line.

5 Write all the numbers that are:

a greater than 200 but less than 210

b less than 190 but greater than 185

c greater than 934 but less than 943.

➡ *Workbook page 12*

Round to the nearest 10

We can **round** numbers to the nearest multiple of 10.

To round to the nearest 10, look at the digit in the ones place.

If it is 4 or less, **round down**. If it is 5 or greater, **round up**.

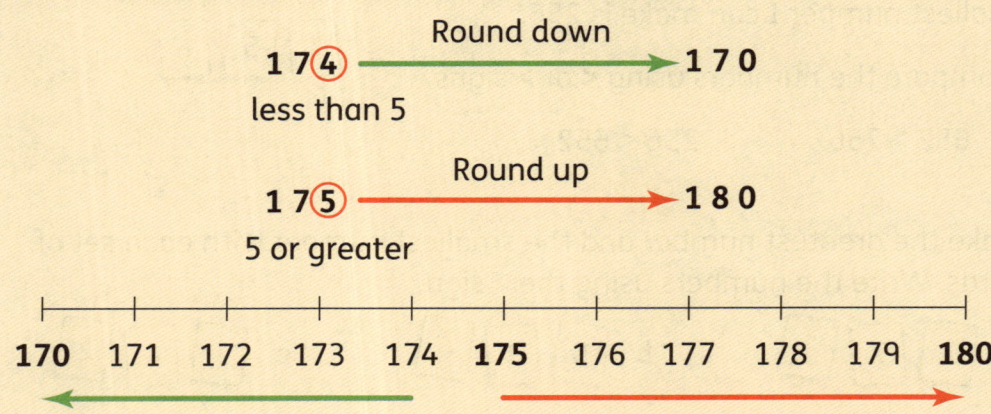

1 **a** Draw a number line from 0 to 10. Circle all the numbers that you would round down to zero.

   b Draw a number line from 10 to 20. Circle all the numbers that you would round up to 20.

2 **a** Nisha rounded a number up to the nearest 10. Her answer was 50. What possible numbers could she have rounded?

   b Kai rounded a number down to the nearest 10. His answer was 170. What possible numbers could he have rounded?

3 Round each number to the nearest 10.

   a 128　　　**b** 374　　　**c** 685　　　**d** 941　　　**e** 288

 Problem solving

> We add numbers together to find the sum. Sum means the same as total.

4 **a** I am a 3-digit number. The **sum** of my digits is 7. When you round me to the nearest 10, I make a number with 5 in the hundreds place and zero in the tens and ones places. What number am I?

   b I am a 3-digit number. When you round me to the nearest 10, I make a 4-digit number. The sum of my digits is 27. What number am I?

   c Make up your own number rounding problem for a partner to solve.

Round to the nearest 100

We can round numbers to the nearest multiple of 100.

To round to the nearest 100, we look at the digit in the tens place.

If it is 4 or less, we round down. If it is 5 or greater, we round up.

All the numbers from 101 to 149 round down to 100.

All the numbers from 150 to 199 round up to 200.

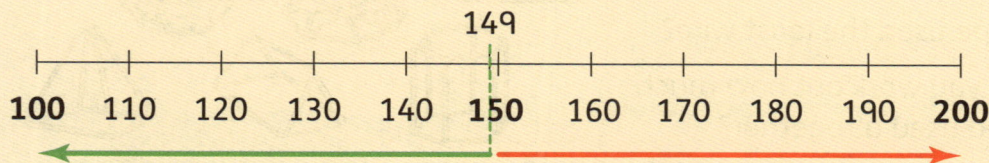

1 Write the answers.

   a Is 189 closer to 100 or to 200?   **b** Is 393 closer to 300 or to 400?

   c Is 529 closer to 500 or to 600?   **d** Is 264 closer to 200 or to 300?

   e Is 345 closer to 300 or to 400?   **f** Is 364 closer to 300 or to 400?

   g Is 751 closer to 700 or to 800?   **h** Is 599 closer to 500 or to 600?

2 Round each number to the nearest 100.

   a 137   **b** 356   **c** 635   **d** 379   **e** 219

   f 799   **g** 890   **h** 850   **i** 670   **j** 909

3 Which numbers in the box round to 400, to the nearest 100?

> 435   496   378   416   351   387   421   395   463   480

4 Which numbers in the box round to 900, to the nearest 100?

> 924   881   838   937   892   945   973   876   915   869   950

➡ *Workbook page 13*

Length

Measuring length

Think and share

Ahmed made some shapes out of wire.

- Which shape used the most wire?
- Which shape used the least wire?
- How could you work out how much wire Ahmed used altogether?

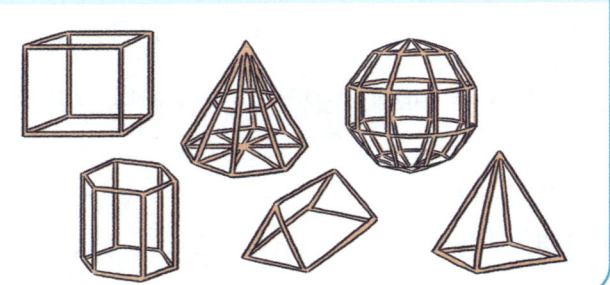

This ruler is marked in **centimetres** (cm). Each centimetre is made up of 10 **millimetres** (10 mm).

When you measure with a ruler, always start measuring from 0.

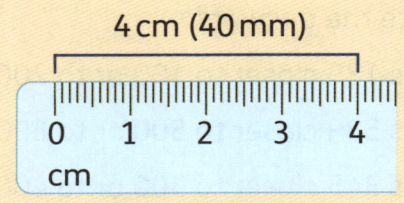

4 cm (40 mm)

1 **a** Estimate how long these objects are in centimetres. Write your estimates.

   b Measure the objects. Write each measurement to the nearest centimetre.

A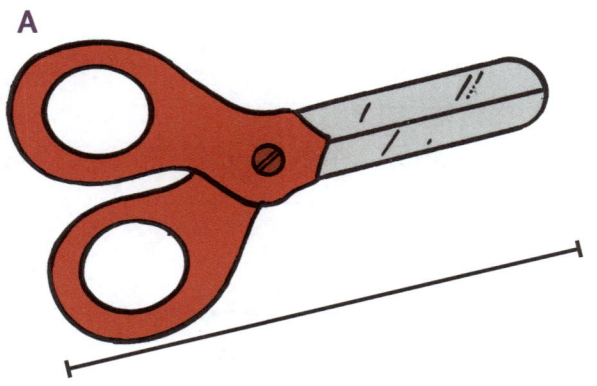

B

C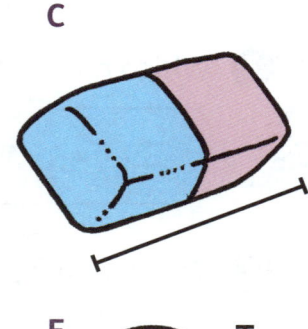

D

E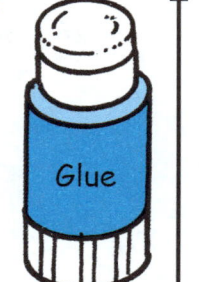

Glue

Measure paths

To measure the **length** of a path:

- measure each part of the path
- write each measurement in centimetres (cm)
- add the measurements to get the total length.

2 cm + 1 cm + 3 cm = 6 cm

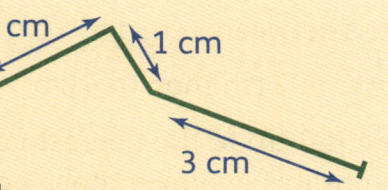

1 How long is each path?

a
b
c

d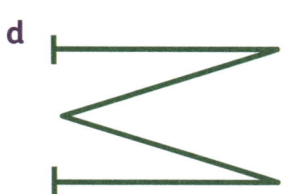
e
f

2 Write the lengths of the paths in order from longest to shortest.

3 Use your ruler and coloured pencils. Measure and draw these paths. Write the total length under each path.

a 6 cm + 1 cm   b 7 cm + 9 cm

c 8 cm + 3 cm + 2 cm   d 4 cm + 5 cm + 3 cm

4 Draw a path. Then measure part of the path to cut off. What length is left?

a 7 cm − 4 cm   b 10 cm − 8 cm

 You can measure each side, then add.

Problem solving

5 Here are three **triangles** made out of wire. Guess which one used the most wire. Then measure to see whether you were correct.

a
b
c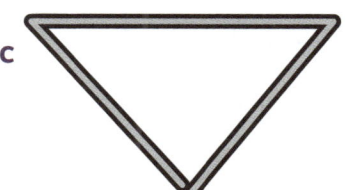

➡ *Workbook page 14*

Round to the nearest centimetre

This line is between 3 cm and 4 cm long.

It is closer to 3 cm than to 4 cm.

We say it is 3 cm to the nearest centimetre.
This means it is approximately 3 cm long.

1 cm = 10 mm

If you count 5 or more millimetres after the centimetre measurement, round up.

If you count 4 or fewer millimetres after the centimetre measurement, round down.

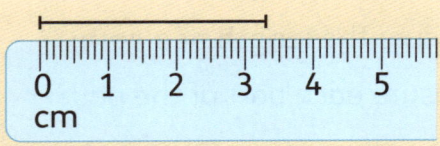

1 Estimate the length of each bead bracelet in centimetres.
Write your estimate.

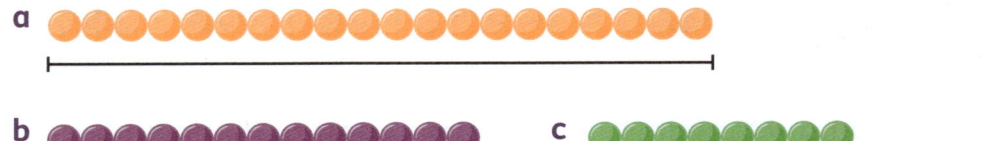

2 Now measure the bracelets. Write each length correct to the nearest centimetre.

3 Write the lengths in order from shortest to longest.

4 Use a ruler. Draw lines that are approximately (not exactly) these lengths.

a 4 cm   b 12 cm   c 8 cm

d 9 cm   e 2 cm   f 5 cm

Metres

You will need a metre stick, a measuring tape or a piece of string that is 1 metre long.

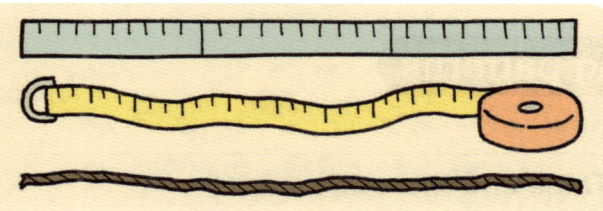

1. Use your **metre** measure to measure each item.

   a

   the length of a rug

   b

   the width of an arm span

   c

   the length of a shoe

   d

   5 X 3 = 15
   the length of the board

   e

   the width of a school desk

   f

   the height of a school chair

2. Copy this table. Tick the right block for each measurement.

	What I measured	Less than 1 m	About 1 m	More than 1 m
a	rug			
b	arm span			
c	shoe			
d	board			
e	school desk			
f	school chair			

3. Discuss with your class.

   a When do people need to estimate in metres in everyday life?

   b How can you become better at estimating in metres?

4. Find out how many metres make 1 kilometre (1 km).
   What things do we measure in kilometres?

▶ *Workbook page 15*

Making patterns

> **Think and share**
>
> Look at the pattern and try to describe it.
>
> What numbers did you use in your description?
>
> What different ways could you continue
> the pattern?

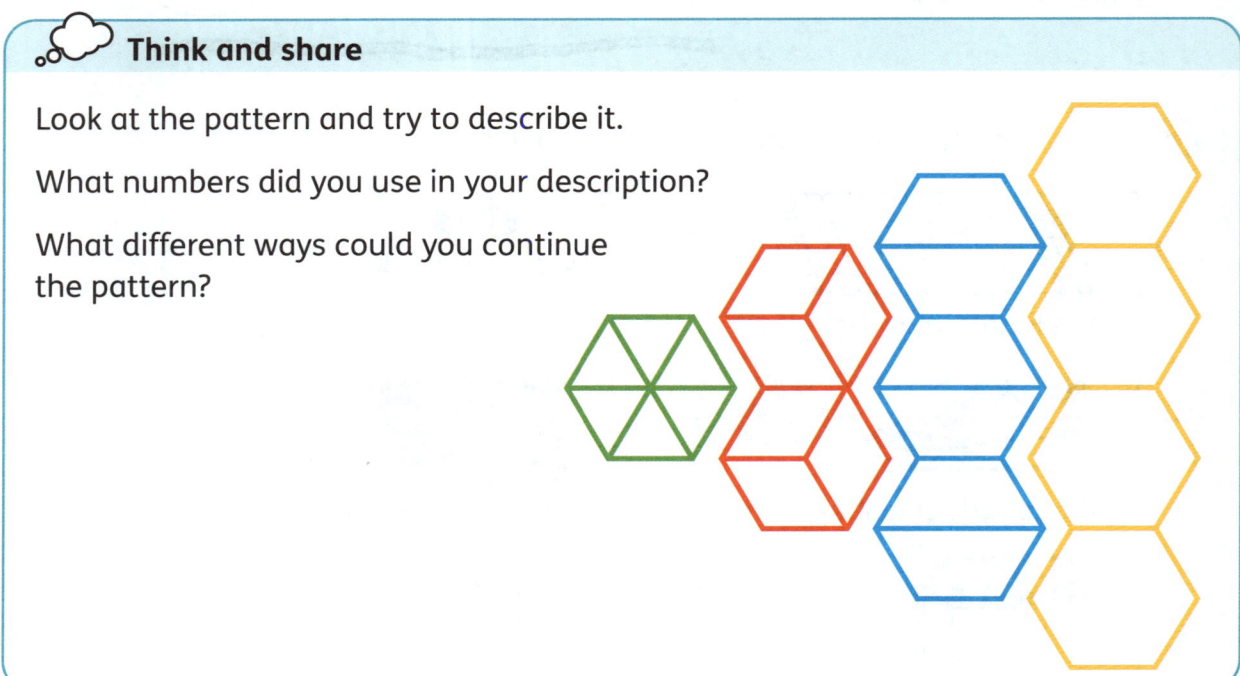

1 Describe each pattern in two ways: using shapes and using numbers.

How would you draw the next shape in each sequence?

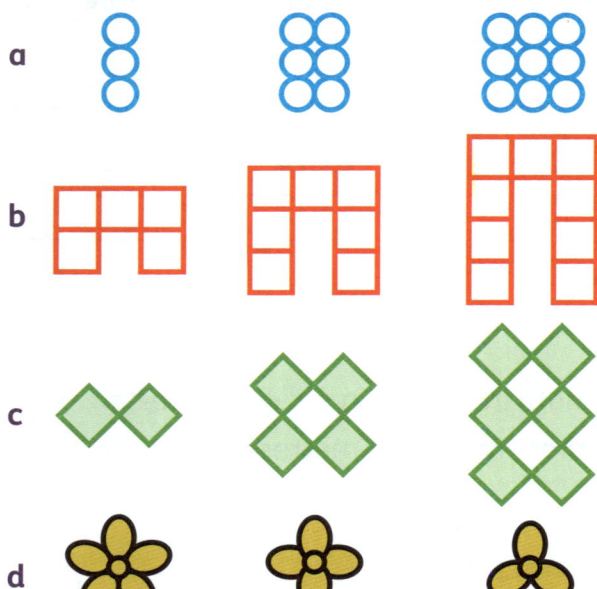

a

b

c

d

➡ *Workbook page 16*

Number sequences

A number pattern is also called a **sequence**. Each number in the sequence is called a **term**. Look at this sequence.

1, 3, 5, 7, 9 …

What kinds of numbers are in this sequence?

What do you do to find the next term in the sequence?

1 You can use the 100 chart to help you count on and back in steps of different sizes.

1	2	3	4	5	6	7	8	9	10
11	12	13	14	15	16	17	18	19	20
21	22	23	24	25	26	27	28	29	30
31	32	33	34	35	36	37	38	39	40
41	42	43	44	45	46	47	48	49	50
51	52	53	54	55	56	57	58	59	60
61	62	63	64	65	66	67	68	69	70
71	72	73	74	75	76	77	78	79	80
81	82	83	84	85	86	87	88	89	90
91	92	93	94	95	96	97	98	99	100

   a What do you notice about the different colours in the square?

   b Count on in twos from 40 to 80.

   c Count on in fives from 10 to 60.

   d Count back in twos from 49 to 29.

   e Count back in fives from 100 to 50.

2 These sequences are made by counting on or back in equal steps. Write the next 3 terms in each sequence.

   a 2 4 6 8 10

   b 25 24 23 22 21

   c 0 3 6 9 12

   d 1 3 5 7 9

   e 15 20 25 30 35

   f 100 90 80 70 60

3 These sequences are made by counting on or back in equal steps. Work out the size of the steps in each sequence. Count on or back to find the next 6 terms.

   a 4 8

   b 33 30

   c 95 85

   d 900 800

   e 0 10

   f 3 10

➡ *Workbook page 17 and page 18*

Term-to-term rules

Look at this number sequence. **19   21   23**   ...

The first term is 19. The second term is 21. The third term is 23.

The **term-to-term rule** is the rule that you use to work out the next term.

The term-to-term rule for this sequence is 'count on in 2s'. What is the next term?

1 Talk about these sequences with a partner. Work out the term-to-term rule for each sequence. What are the next 3 terms?

a 45 50 55

b 85 80 75

c 27 30 33

d 24 28 32

2 Read these instructions for some sequences. Write the first 6 terms in each sequence.

a Start at 20 and count on in tens.

b Start at 35 and count on in fives.

c Start at 19 and count on in threes.

d Start at 50 and count back in fours.

e Start at 45 and count back in fives.

f Start at 45 and count back in threes.

3 Tell your partner the first term in each sequence and the term-to-term rule.

a 25 30 35 40

b 26 36 46 56

c 89 79 69 59

d 3 13 23 33

e 34 36 38 40

f 47 50 53 56

4 Play with a partner. Take turns to pick two digit cards, from 1 to 9.

Write the 2-digit numbers that you can make with the digits.

- Write the number that is 5 less than each number.
- Write the number that is 5 more than each number.
- Write the number that is 10 more than each number.
- Write the number that is 10 less than each number.

➡ *Workbook page 19*

Using shapes and objects

In additions and subtractions, we sometimes use a shape or picture to represent an unknown number. Look at this example. Four pupils worked out the unknown number in different ways.

$19 - \triangle = 7$

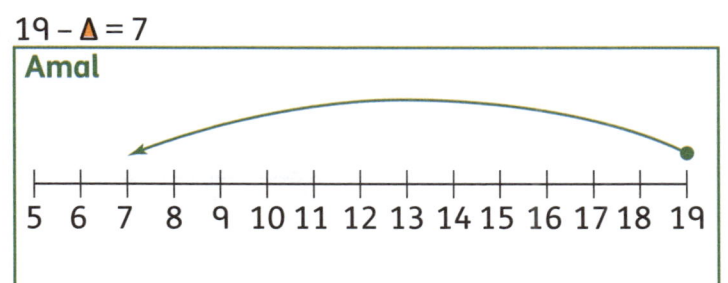

Amal

Jip

$19 - 9 = 10$
$10 - 3 = 7$
I took away 9.
I took away another 3.
$9 + 3 = 12$
$19 - 12 = 7$

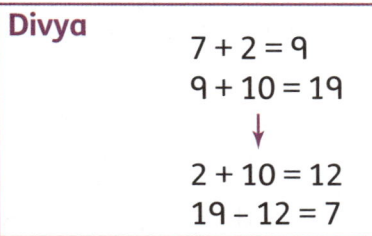

Divya

$7 + 2 = 9$
$9 + 10 = 19$

$2 + 10 = 12$
$19 - 12 = 7$

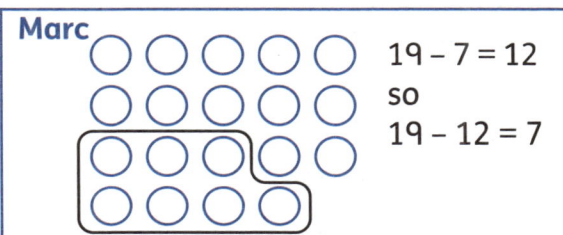

Marc

$19 - 7 = 12$
so
$19 - 12 = 7$

1 Discuss with a partner why all these ways work.

2 Use any method to work out what each shape represents.

a $178 + $  $= 179$

b $350 - $ 🍃 $= 340$

c $500 + $ 🌼 $= 600$

d ⚡ $- 100 = 900$

3 The rule is ⭐ = 💚 + 1.

Work out these values. The first one is done for you.

a If 💚 = 4, then ⭐ = 5

b If 💚 = 7 then, ⭐ = ☐

c If 💚 = 20, then ⭐ = ☐

d If ⭐ = 18, then 💚 = ☐

💡 Problem solving

 What number facts could help you?

e ⭐ + 💚 = 21

Using the same rule, find the values of the shapes.

UNIT 5 Lines and angles

Angles

Think and share

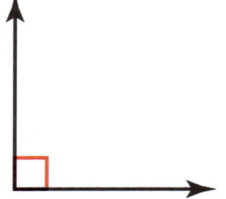 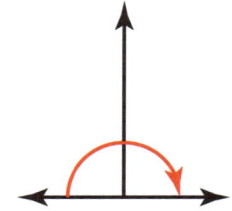 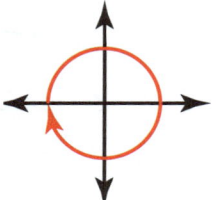

A square corner is called a **right angle**.

Two right angles make a **half turn** or a straight line.

Four right angles make a **full turn**.

Look at this pattern.

- Which angles are right angles?
- Which angles are larger than right angles?

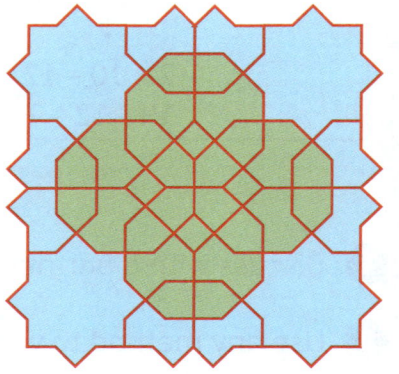

1 Look at the angles in these sets of shapes.

Which shape is the odd one out? Explain why.

a

b

c A

Right angles

A right angle is also called a **square corner** or a **90-degree angle**.

The little square shows that it is a right angle.

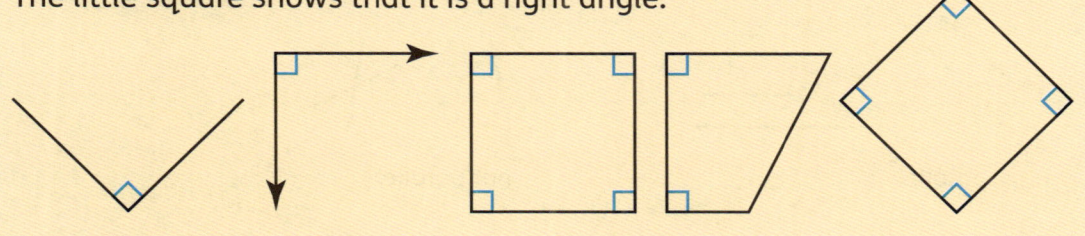

1 How many right angles can you find in each shape?

a   b   c   d

  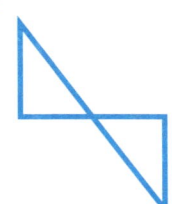 

2 Which of these triangles have a right angle?

A   B   C   D   E

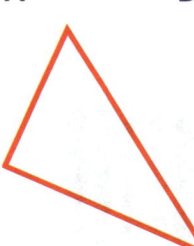 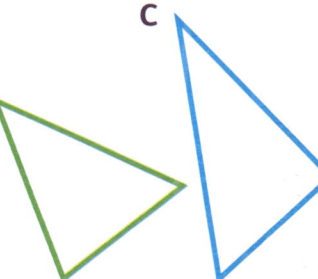 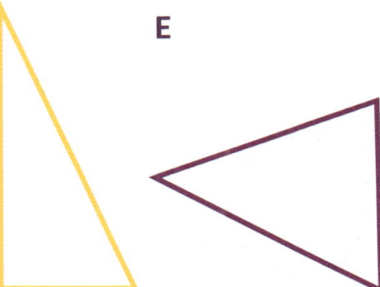

3 True or false? A triangle always has a right angle.

4 Is it possible to draw a triangle that has two right angles? Work with a partner. Give reasons or draw sketches to explain your answer.

Problem solving

5 Can you draw a shape with 10 right angles?

> Sketch some ideas on paper. Think about things you see in real life that have right angles.

➡ *Workbook page 20*

Parallel and perpendicular lines

Parallel lines stay the same distance apart. They never meet.

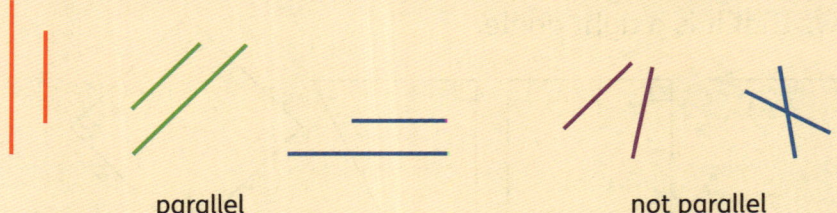

parallel   not parallel

Perpendicular lines meet at a right angle.

perpendicular   not perpendicular

1 Which set of lines is *not* parallel?

A   B   C   D

2 Which of these pictures show parallel lines?

A   B   C

3 Draw two pairs of parallel lines.

4 Identify a pair of perpendicular lines in each picture. Draw and label them.

a   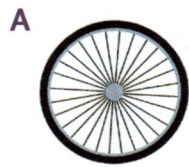   b      c

5 **a** Find three things in your classroom that have perpendicular lines.

   b Find three things that have parallel lines.

Horizontal and vertical

The horizon is the line where the Earth appears to meet the sky.

Horizontal lines run from side to side, like the horizon.

Vertical lines run straight up and down.

Horizontal lines are perpendicular to vertical lines.

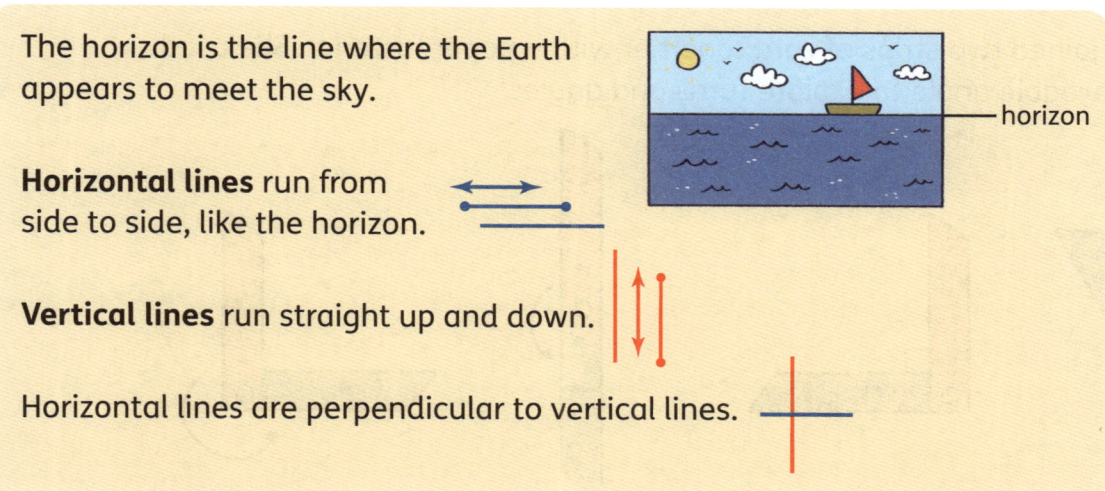
horizon

1 Look at each object. Is it horizontal or vertical?

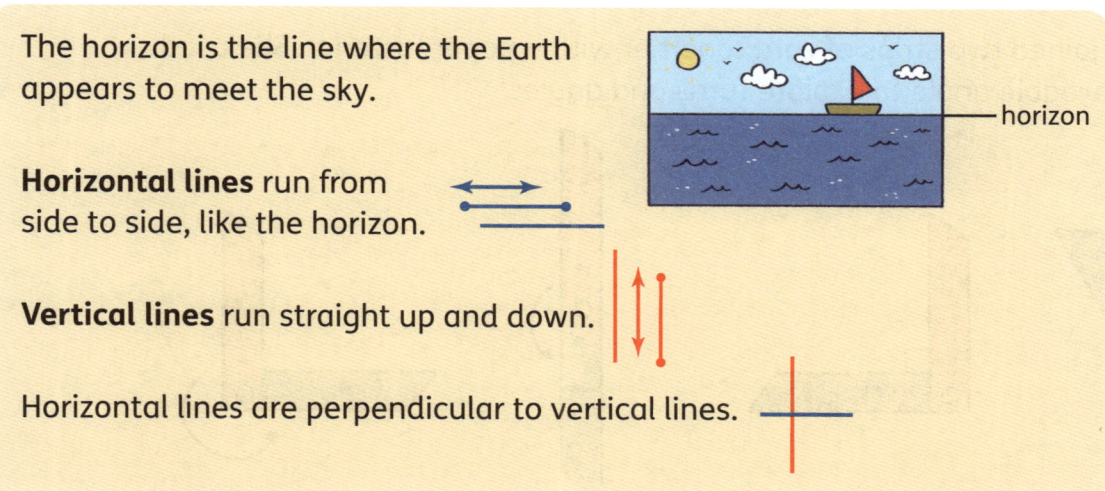

a   b   c   d

e   f   g   h

2 Look at the hands on each clock. Is each hand horizontal, vertical or neither?

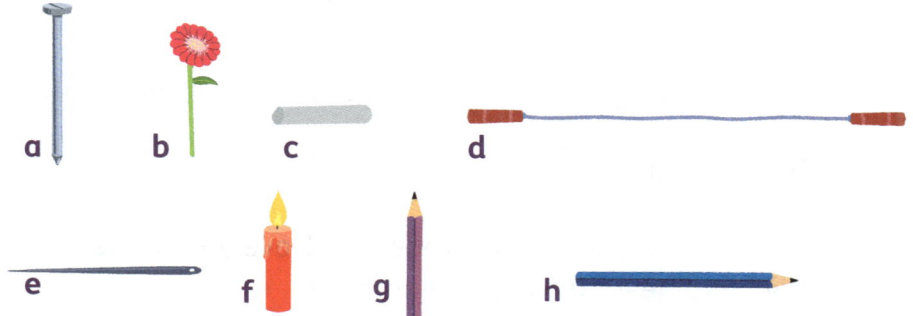

a   b   c   d

3 Discuss these patterns with a partner. Describe what you see. Identify horizontal, vertical, parallel and perpendicular lines.

a   b   c

Turns

Emma joined two strips of card together with a paper fastener. She used this moveable angle to explore turns and angles.

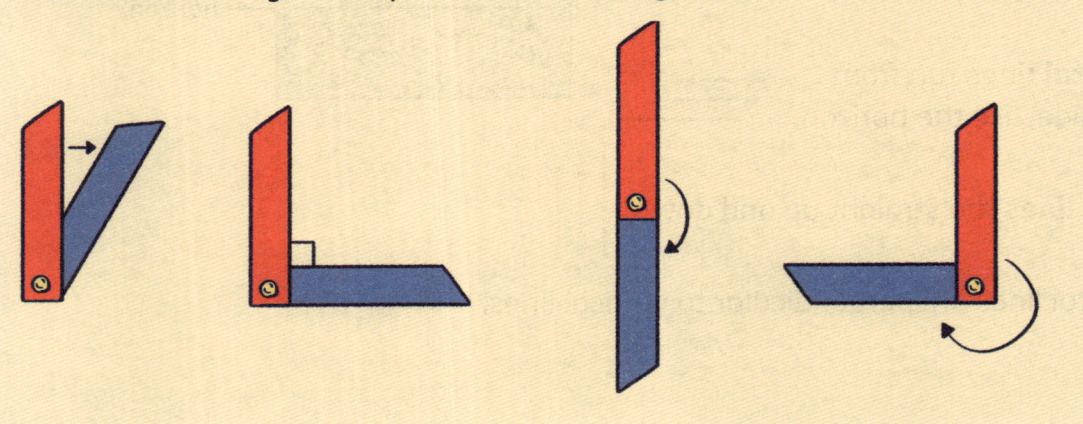

1. Use strips of card and a paper fastener to make your own moveable angle. Make:

   a a right angle   **b** a straight line   **c** a $\frac{3}{4}$ turn   **d** a full turn.

2. What happens when you make a full turn?

3. Work with a partner. Put two right angles together. Have you made parallel or perpendicular lines? Can you explain why?

4. Look at each turn. Is it a **quarter turn**, a half turn or a **three-quarter turn**?

   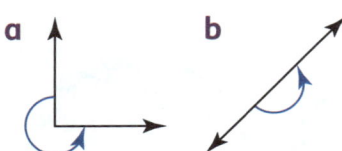      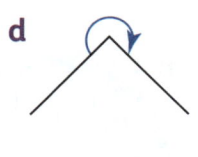   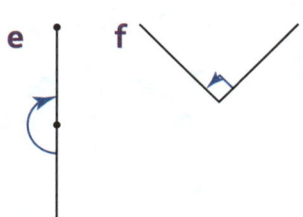

5. Look at each angle. Is it less than, equal to or more than a right angle?

   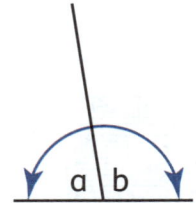   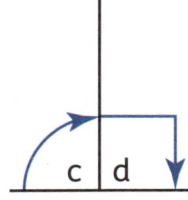   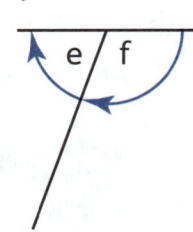

➡ *Workbook page 21*

Polygons

Polygons

Think and share

Samira is an artist who creates patterns using geometric shapes.

What shapes do you see? What angles do you notice?

Do you remember the names of these polygons?

square　　　　**rectangle**　　　　**triangle**　　　　**pentagon**　　　　**hexagon**

'Poly' means many and '-gon' means angle. A **polygon** is a shape that has straight sides. The corners of a polygon are called **vertices**. One corner is a **vertex**.

1 Which polygon am I? Match each description to the correct polygon name.

　a I have 3 sides and 3 vertices.

　b I have 5 sides and 5 vertices.

　c I have 6 vertices. If you fold me in **half**, I have 4 sides.

　d I have 4 sides that are all equal in length.

Problem solving

 Make drawings or paper cut-outs to help you.

2 Which shapes can you fold in half to make right-angled triangles?

Regular and irregular polygons

Regular polygons have sides that are all the same length and inside angles that are all the same size. Irregular polygons have sides of different lengths.

Polygon	Number of sides	Examples
Triangle	3	
Quadrilateral	4	
Pentagon	5	
Hexagon	6	
Octagon	8	

1. Count sides and write the name of each polygon. Is it regular or irregular?

    a   b   c   d   e

2. Which polygon in each set is the odd one out?

a   

b   

3. Use a ruler. Draw four irregular and four regular polygons. Swap with a partner. Count the sides and name each polygon.

➡ *Workbook page 22*

Angles in shapes

We can cut out a **circle** and fold it in **quarters** to make a right angle.
We can use this right angle to measure the angles inside shapes.

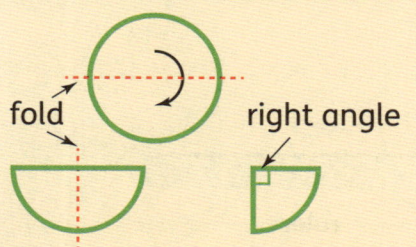

fold

right angle

We measure angles in **degrees**.

A right angle is equal to
90 degrees. We write 90°.

1 Make a right angle. Use your right angle to measure the angles in these shapes. Write the letters of:

   a right angles   **b** angles greater than 90°   **c** angles less than 90°.

2 Measure the angles in these shapes. Write the letter of each angle.
Then write < 90°, 90° or > 90° next to the letter.

  a
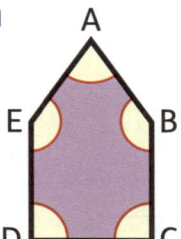
  b
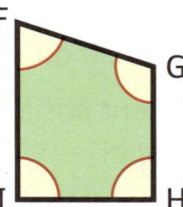
  c
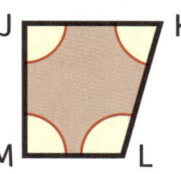
  d
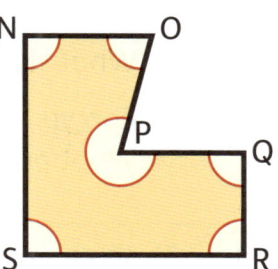

3 Look at the angle between the hands on each clock.
Write < 90°, 90° or > 90°.

  a
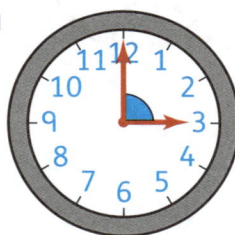
  b

  c
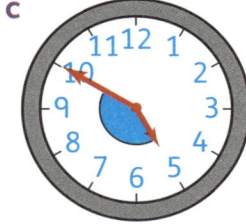
  d

Sketch 2D shapes

We can sketch 2D shapes using some special instruments.

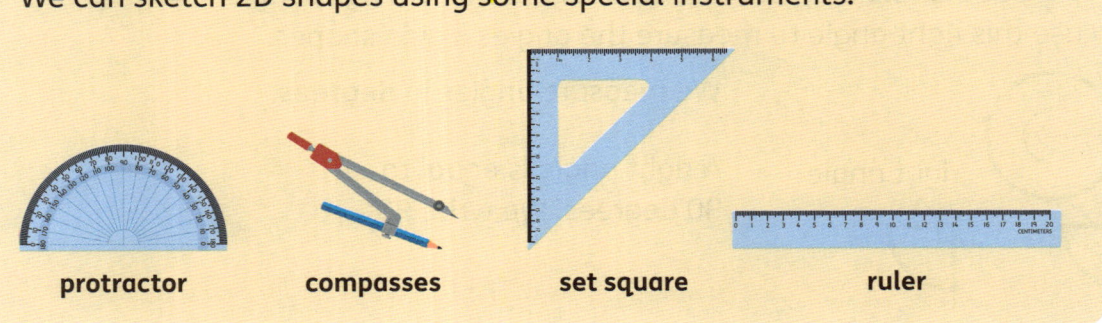

protractor   compasses   set square   ruler

1 Follow these instructions.

- Draw a horizontal line. Mark a point on your line and label it A.

- Measure 3 cm from point A. Label this point B.

- Draw perpendicular lines at points A and B. Use a set square.

- Measure 3 cm up from point A. Label this point D. Measure 3 cm up from point B. Label this point C.

a Which pairs of lines are perpendicular? Which pair of lines is parallel?

b What shape will you make if you join points C and D?

c What shape will you make if you join points D and B?

d Can you make the same shape in a different position? What points do you join to make this shape?

Problem solving

2 Look at these overlapping circles. How can you use circles like these to help you draw triangles?

➡ *Workbook page 23*

Symmetry

Here is a square piece of paper.

The line shows where the paper has been folded in half.

The fold line is a **line of symmetry.**

When we fold a shape along a line of symmetry, the two parts match.

line of symmetry

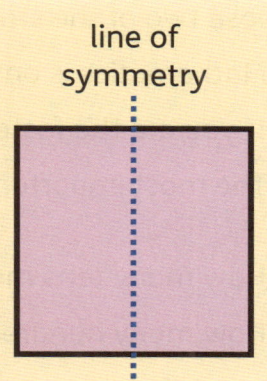

1 Cut out a square piece of paper. How many different ways can you fold the square to make two matching parts?

2 Look at these shapes. How many different ways can you fold each shape in half?

a    b    c    d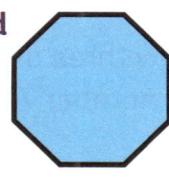

3 Fold a piece of paper and cut out a shape like this. What happens when you open the shape?

4 How many lines of symmetry does this shape have?

 Problem solving

5 Work with a partner. Try to cut out a shape with more than one line of symmetry.

➤ Workbook page 24 and page 25

Mixed practice 1

1 Choose two of the statements and explain why they are incorrect.

- Mathematics is only about numbers, not about real life.
- If you are the fastest, you are the best at maths.
- The most important thing in maths is to get every answer right.

2 **a** How many tens make 100?

b How many hundreds make 1000?

3 What numbers are missing from this number line?

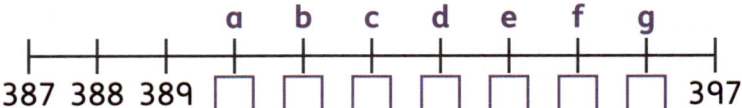

4 Use the digits 8, 1 and 5. Make four 3-digit numbers and four 2-digit numbers. Then write them in order from smallest to greatest.

5 Show three different ways to partition the number 749. Use diagrams like these.

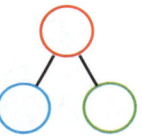 

6 Round each number to the nearest 10.

a 451   **b** 378   **c** 605   **d** 844

7 Round each number to the nearest 100.

a 182   **b** 257   **c** 448   **d** 974

8 Measure the length of each leaf. Write each length to the nearest cm.

9 Draw a triangle that has one side 5 cm long and another side 3 cm long. Draw the third side and measure it correct to the nearest cm.

10 Draw the next shape in this sequence. Write the number sequence for this pattern.

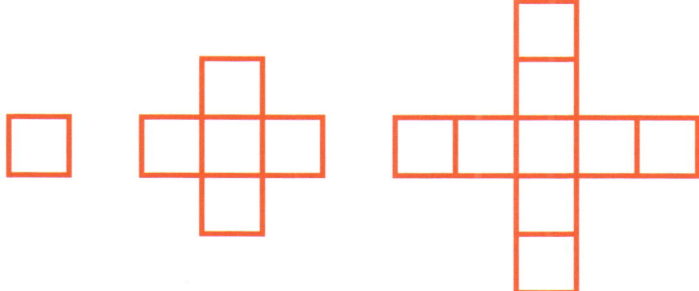

11 **a** Draw a shape sequence that matches this number sequence: 2 4 6 8.

b Write two different ways to describe the rule for this number sequence.

12 **a** What is another name for a square corner or a quarter turn?

b How many quarter turns make a full turn?

13 Write the letters of the shapes that are:

a hexagons

b triangles

c pentagons

d regular polygons.

Write the letters of the shapes that have:

e parallel lines

f perpendicular lines.

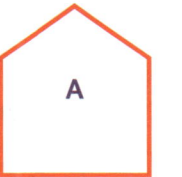
A

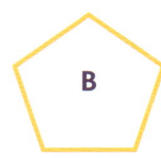
B

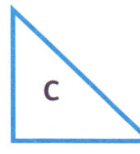
C

E

D

F

G

14 Explain in your own words what parallel and perpendicular lines are. Draw a sketch to show each type of line.

Addition and subtraction

Number facts and families

> **Think and share**
>
> Each hole in these number frames represents 1.
>
> How can you work out the total that these number frames represent?
>
> Work with a partner. Then compare your working with other pairs. What different ways did each pair find?
>
> Now imagine each hole represents 10. What total do these number frames represent?

1 Look at these number frames and **fact families**.

 $10 + 0 = 10$

$10 - 0 = 10$

 $8 + 2 = 10$ $10 - 2 = 8$

$2 + 8 = 10$ $10 - 8 = 2$

   a Draw a picture with two number frames that make 10. Write the fact family for your picture.

   b How can fact families help us to check addition and subtraction?

2 Work with a partner.

   a Use the number frames to make 10 in different ways. Say the fact family for each way.

   b Now each hole represents 10. So, one dark blue number frame represents 100. Use the number frames to make 100 in different ways. Say the fact family for each way.

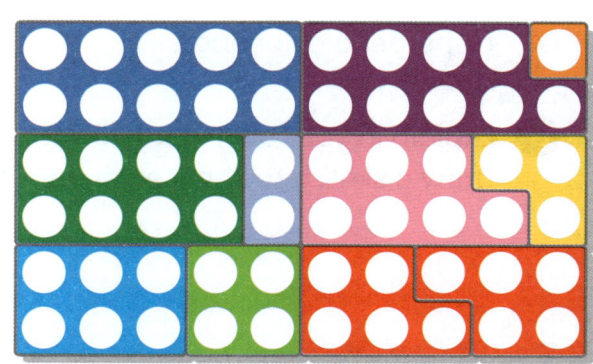

   c Now each hole represents 100. What fact families can you make?

Addition and subtraction facts

Look at some different **strategies** that you can use for adding and subtracting. A strategy is a plan for working something out.

$4 + 6 = \square$	I know the answer by heart. $4 + 6 = 10$
$12 + \square = 20$	I know that $2 + 8 = 10$, so $12 + 8$ is 10 more.
$2 + 14 = \square$	I re-order to make it easier to count on. 14, 15, 16
$9 + 8 = \square$	I know that double 8 is 16 and the next number is 17.
$15 - 7 = \square$	I know $7 = 5 + 2$. So I take away 5 from 15 to make 10. Then $10 - 2 = 8$.
$17 - 8 = \square$	I can make this subtraction easier. I know that $17 - 7 = 10$. Then I take away 1 more. $10 - 1 = 9$

1 Copy the number sentences. Fill in the missing numbers.

a $7 + 3 = \square$ b $\square - 7 = 3$ c $13 + \square = 20$

d $15 + \square = 20$ e $20 - 15 = \square$ f $20 - 5 = \square$

g $19 + 1 = \square$ h $1 + \square = 20$ i $20 - 19 = \square$

2 Use the three numbers in each box to write four different number sentences.

For example: | 13   7   20 |

$13 + 7 = 20$ $7 + 13 = 20$ $20 - 13 = 7$ $20 - 7 = 13$

a | 15   2   17 | b | 13   4   17 |

c | 17   3   14 | d | 18   3   15 |

e | 13   18   5 | f | 7   6   13 |

➡ *Workbook page 26*

Addition and subtraction problems

Use cubes, counters or number frames if you need them to help you work out the answers.

1 Copy the additions. With a partner, talk about which facts you could use to make each addition easy to solve. Write the missing numbers.

a ☐ + 6 = 13   ☐ + 60 = 130   b ☐ + 9 = 15   ☐ + 90 = 150

c ☐ + 4 = 12   ☐ + 40 = 120   d 5 + ☐ = 14   50 + ☐ = 140

e 3 + ☐ = 11   30 + ☐ = 110   f 8 + ☐ = 17   80 + ☐ = 170

💡 **Problem solving**

2 Write a number sentence for each problem. Work out the answers.

a A football team needs 11 players. It has only 3 players. How many more players does the team need?

b Ranjit has 5 marbles. He buys 8 more marbles. How many marbles does Ranjit have altogether?

c A gardener planted 15 seeds. Only 8 seeds grew. How many seeds did not grow?

d Sally scored 16 points. Ricky scored 9 points. What is the difference?

e There are 18 crayons in a tub. 9 pupils each take one crayon. How many crayons are left?

f Busi needs 8 more counters to make 14. How many counters does she have already?

➡ *Workbook page 27*

Addition and subtraction patterns

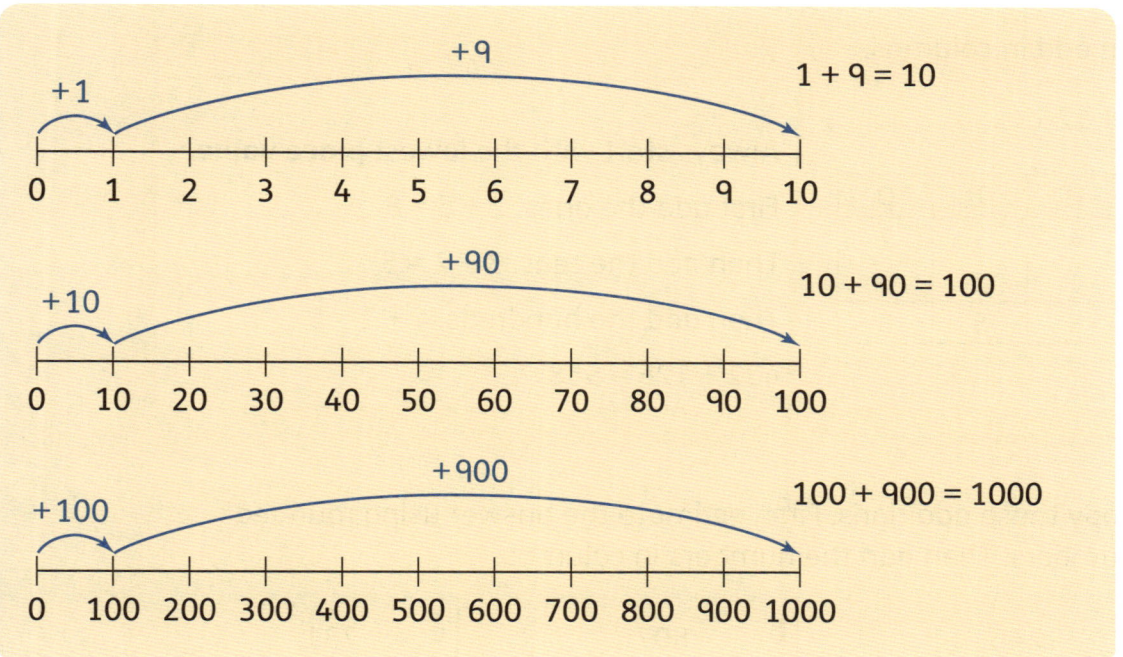

1 Copy and complete the number sentences.

a $8 + 2 = \square$ **b** $80 + 20 = \square$ **c** $800 + 200 = \square$

d $3 + \square = 10$ **e** $30 + \square = 100$ **f** $300 + \square = 1000$

2 Copy and complete the number sentences.

a $1000 - 100 = \square$ **b** $1000 - 500 = \square$ **c** $1000 - 400 = \square$

d $1000 - \square = 200$ **e** $1000 - \square = 700$ **f** $\square - 1000 = 0$

3 How much water must we add to each jug to fill it to 1000 ml?

First read how much water is in the jug. Then work out how much more water we need to make 1000 ml.

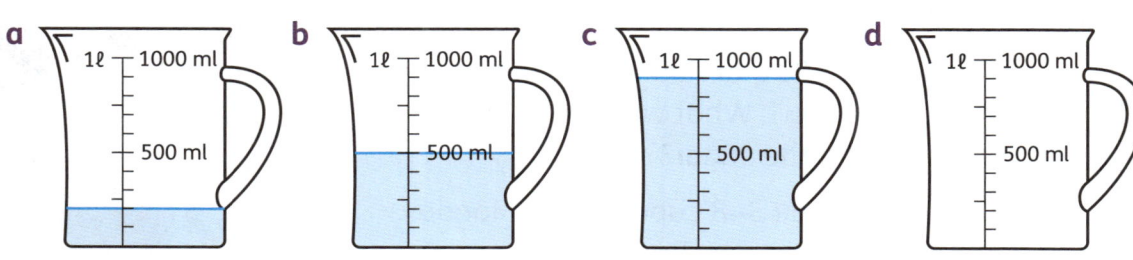

Workbook page 28

Work in columns

We can add in columns.

432 + 126

H	T	O
4	3	2
+ 1	2	6
5	5	8

Always start with the lowest **place value**.

First add the ones. 2 + 6 = 8

Then add the tens. 3 + 2 = 5

Then add the hundreds. 4 + 1 = 5

432 + 126 = 558

1 Copy these additions. First, estimate the answer using rounded numbers. Then add the numbers in columns.

a
```
   246
 + 613
 —————
```

b
```
   807
 +  41
 —————
```

c
```
   731
 + 220
 —————
```

d
```
   480
 +  15
 —————
```

e
```
   106
 + 101
 —————
```

f
```
   111
 + 222
 —————
```

g
```
   171
 +  26
 —————
```

h
```
   305
 + 114
 —————
```

i
```
   371
 + 423
 —————
```

 Problem solving

2 Solve these problems.

a A baker put 250 g of flour and 235 g of sugar in a mixing bowl. What was the total mass of the flour and the sugar?

b The baker made 348 cupcakes on Monday and 121 cupcakes on Tuesday. How many cupcakes did she make altogether?

lesson continues ◗

We can also **subtract** in columns.

714 – 302

H	T	O
7	1	4
– 3	0	2
4	1	2

Subtract the bottom number from the top number.

Always start with the lowest place value.

Subtract the ones. $4 - 2 = 2$

Subtract the tens. $1 - 0 = 1$

Subtract the hundreds. $7 - 3 = 4$

$714 - 302 = 412$

3 Copy and complete these subtractions.

a
```
   873
 - 612
 -----
```

b
```
   349
 - 217
 -----
```

c
```
   855
 -  34
 -----
```

d
```
   791
 - 501
 -----
```

e
```
   787
 -  66
 -----
```

f
```
   122
 - 111
 -----
```

g
```
   417
 - 205
 -----
```

h
```
   583
 - 422
 -----
```

i
```
   830
 - 400
 -----
```

4 a Start with 500. Take away 2 hundreds. How much is left?

b Start with 640. Take away 2 tens and 2 hundreds. How much is left?

c Start with 995. Take away 10 ones. How much is left?

d Start with 473. Take away 14 tens. How much is left?

5 Kari made some mistakes in her subtraction problems. Explain to a partner how you would help Kari better understand how to subtract.

```
   1 7 8
 - 1 4
 -------
   1 3 8
```

```
   9 0 5
 -     5 0 0
 -----------
   9 0 0
```

Workbook page 29

Add or subtract multiples of 10 and 100

You can count on or back to add or subtract multiples of 10 and 100.

47 + 50   50 is 5 tens.

The answer is 97.

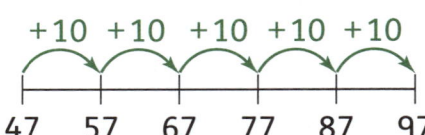

93 – 40   40 is 4 tens.

The answer is 53.

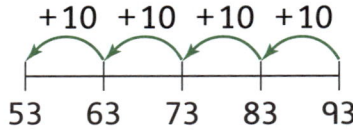

147 + 300   300 is 3 hundreds.

The answer is 447.

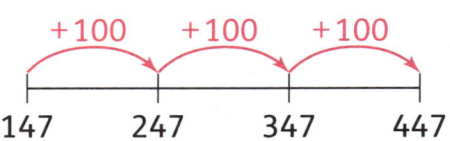

651 – 300   300 is 3 hundreds.

The answer is 351.

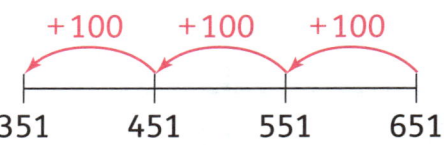

You can also use place value and re-ordering to add or subtract multiples of 10 and 100.

87 + 40   80 + 7 + 40 = 80 + 40 + 7

8 tens plus 4 tens make 12 tens ⟶ 120 ⟶ 120 + 7 = 127

134 + 400   100 + 34 + 400 = 100 + 400 + 34

1 hundred plus 4 hundreds make 5 hundreds ⟶ 500

500 + 34 = 534

1 Add. Then use subtraction to check your answers.

   a 23 + 10   **b** 37 + 60   **c** 94 + 10   **d** 51 + 60

   e 40 + 48   **f** 30 + 91   **g** 124 + 30   **h** 80 + 224

   i 90 + 365   **j** 342 + 100   **k** 254 + 60   **l** 803 + 80

2 Subtract. Then use addition to check your answers.

   a 187 – 10   **b** 342 – 90   **c** 504 – 100

   d 421 – 300   **e** 876 – 50   **f** 420 – 200

More adding and subtracting

$22 + 47 = 69$

Tens	Ones

Add the ones.

Add the tens.

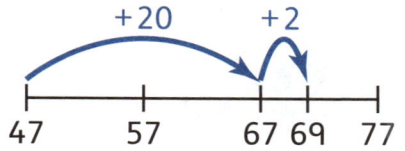

Start with the greatest number.

Count on in groups.

Partition if it helps.

$52 - 34 = 18$

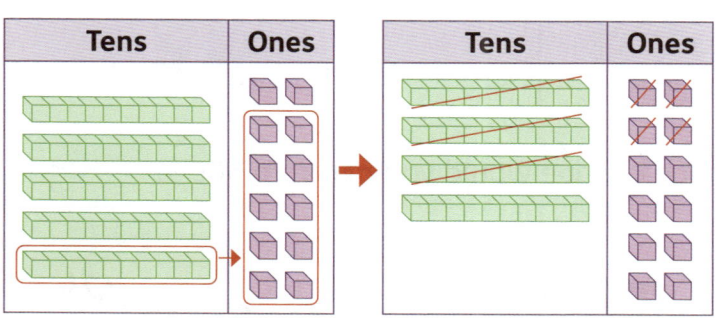

Tens	Ones		Tens	Ones

Regroup the tens and ones.

Next, subtract the ones.

Then, subtract the tens.

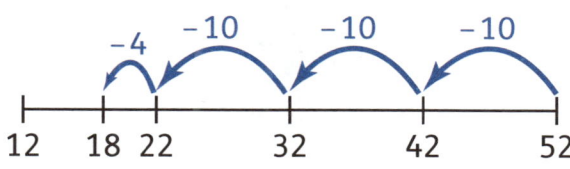

1 Estimate the answer to each calculation first.
Then work it out.
Check your answer afterwards.

a 23 + 45	**b** 44 – 12	**c** 37 + 26
d 63 – 25	**e** 68 – 23	**f** 24 + 47

You can use a number line or place-value table if it helps.

lesson continues ▶

2 Write additions. Work out how much each pair of items costs.

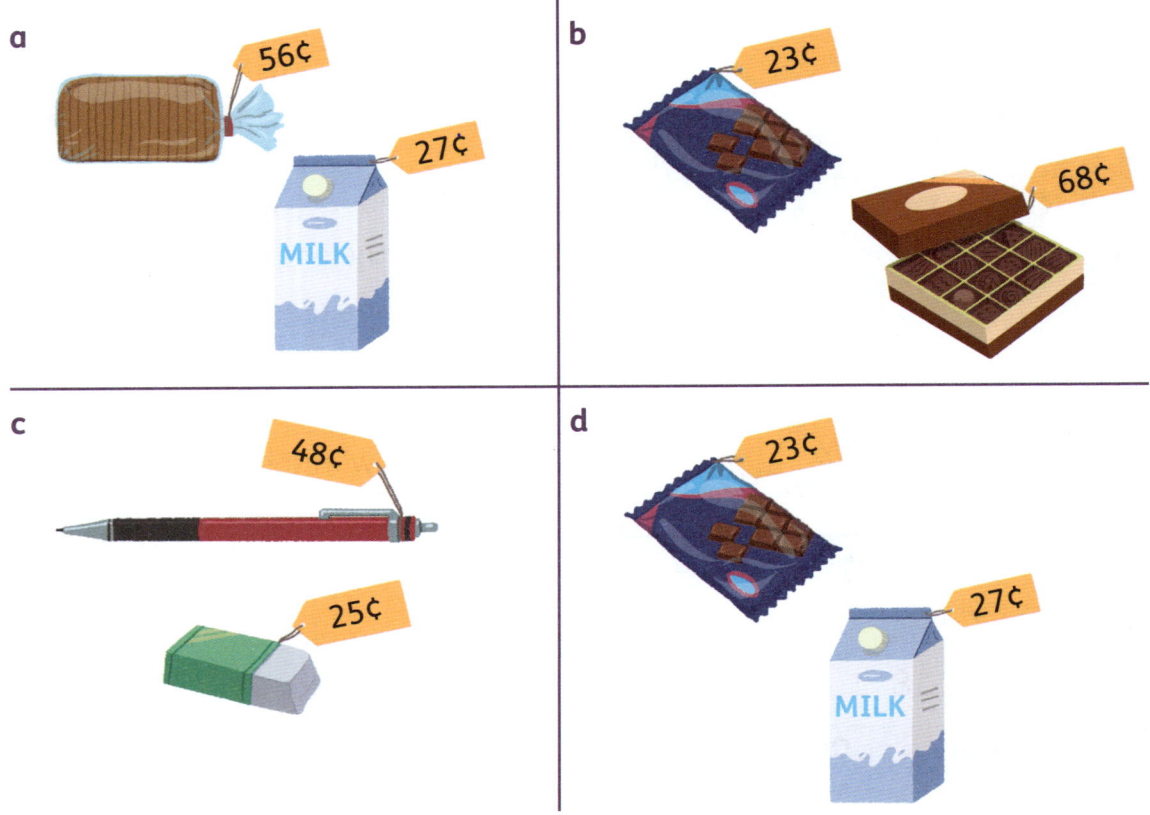

a 56¢ 27¢ MILK

b 23¢ 68¢

c 48¢ 25¢

d 23¢ 27¢ MILK

3 Write subtractions. How much change would you get from 50c if you bought each item?

a 18¢ CEREAL

b SUPER HERO 25¢

c 42¢

4 Four friends go to the beach. Each one has a beach towel.

- The longest towel is 45 cm longer than the shortest towel.
- The shortest towel is 130 cm in length.
- The other two towels are equal in length.
- If they put all the towels in a long row, the total length is 595 cm.

Work out the length of each towel.

➤ *Workbook page 30*

Addition with regrouping

What happens when the digits in a column add up to more than 9?

477 + 215

H	T	O
	1	
4	7	7
2	1	5
6	9	2

+

Ones: 7 + 5 = 12 ones

12 = 1 ten and 2 ones

Write the 2 in the ones column.

Carry the 1 ten to the tens column.

Tens: 1 + 7 + 1 = 9 tens

Hundreds: 4 + 2 = 6 hundreds

477 + 215 = 692

748 + 199

H	T	O
1	1	
7	4	8
1	9	9
9	4	7

+

Ones: 8 + 9 = 17   1 ten and 7 ones
Carry 1 ten.

Tens: 1 + 4 + 9 = 14 tens

14 tens = 1 hundred and 4 tens
Carry 1 hundred.

Hundreds: 1 + 7 + 1 = 9 hundreds

748 + 199 = 947

1 Copy these additions. Estimate each answer first. Then add.
Regroup when you need to.

Use subtraction or any other strategy to check each answer.

a		b		c	
143		208		376	
+ 28		+ 304		+ 192	

d		e		f	
99		48		329	
+ 77		+ 55		+ 166	

g		h		i	
549		107		510	
+ 299		+ 199		+ 190	

➡ *Workbook page 31*

Subtract in columns

$44 - 23 = 21$

H	T	O
	4	4
−	2	3
	2	1

Line up the place values correctly.

Ones: $4 - 3 = 1$

Tens: 4 tens − 2 tens = 2 tens

2 tens and 1 one = 21

$44 - 23 = 21$

$44 - 27 = 17$

H	T	O
	³4̶	¹4
−	2	7
	1	7

Ones: 4 is less than 7, so go to the tens column.
Regroup the 4 tens to make 3 tens and 10 ones.
Now you have 14 ones.

Ones: $14 - 7 = 7$ ones

Tens: $3 - 2 = 1$ ten

1 ten and 7 ones = 17

$44 - 27 = 17$

1 Copy these subtractions and work them out.

a
```
    38
−   14
_____
```

b
```
    59
−   35
_____
```

c
```
    62
−   51
_____
```

d
```
   144
−  122
_____
```

e
```
   461
−  311
_____
```

f
```
   338
−  225
_____
```

2 Copy these subtractions. Use regrouping to help you subtract.

a
```
   475
−  138
_____
```

b
```
   293
−  147
_____
```

c
```
   674
−  199
_____
```

d
```
   850
−  133
_____
```

e
```
   629
−  182
_____
```

f
```
   561
−  177
_____
```

 Problem solving

3 Marco has $873 in his bank account. He spends $385 on a new laptop. How much does he have left?

 Work in columns if it helps.

➡ *Workbook page 32 and page 33*

Part–part–whole

When you add two numbers to make a new number, the new number is called the sum. When we subtract, we take a part away from the whole, and find what is left. The part that is left is called the difference.

Look at this example:

147	
125	22

whole	
part	part

$125 + 22 = 147$
$22 + 125 = 147$

$147 = 125 + 22$
$147 = 22 + 125$

$147 - 125 = 22$
$147 - 22 = 125$

$22 = 147 - 125$
$125 = 147 - 22$

1 Look at the addition $235 + 25$.

   a Work out the sum.

   b Use the same three numbers to write a subtraction, like this:

     □ – □ = □

     Find two different subtractions.

   c What is the difference between 480 and 50?

2 Find the missing part.

   a $124 + \square = 235$ **b** $746 + \square = 859$ **c** $\square + 236 = 547$

   d $468 - \square = 150$ **e** $397 - \square = 236$ **f** $211 - \square = 110$

3 Find the missing whole.

   a $\square - 432 = 102$ **b** $\square = 170 + 30$ **c** $\square - 220 = 220$

Start with some guesses.

💡 Problem solving

4 Emma is 12 years older than her brother and 25 years younger than her father. The sum of all their ages is 67. Can you work out the ages of Emma, her father and her brother?

Money

Write money amounts

> **Think and share**
>
> What kinds of coins and notes do you use in your country?
>
>

Every country has its own coins and notes. Some countries use dollars and cents. There are 100 cents in 1 dollar. 100c = $1.00

We write money amounts with a dot called a **decimal point** between the dollars and cents. We write four dollars and 25 cents as $4.25

When there are no cents, we write zeros as **placeholders**.

For example, we write five dollars as $5.00

$4.25 = 425 cents $5.00 = 500 cents

1   Write these money amounts in numbers.

    a seventy-five cents

    b four dollars and fifty cents

    c three dollars and six cents

    d ten dollars and ninety-nine cents

 Use only the $ symbol.

lesson continues ▶

2 Write the prices in order from cheapest to most expensive.

a

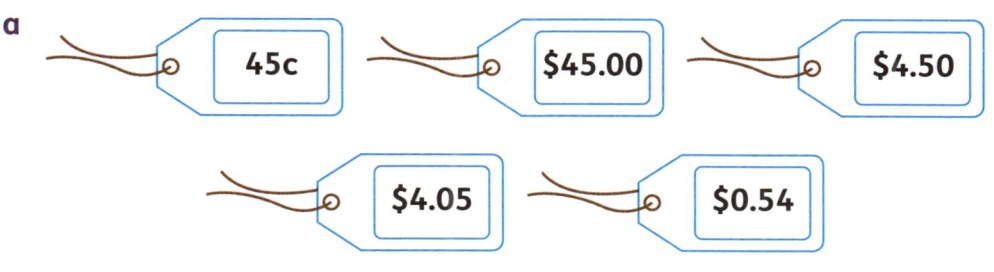

| 45c | $45.00 | $4.50 |

| $4.05 | $0.54 |

b

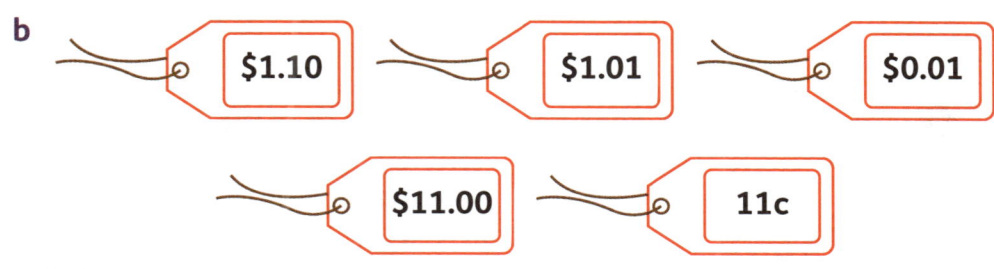

| $1.10 | $1.01 | $0.01 |

| $11.00 | 11c |

c

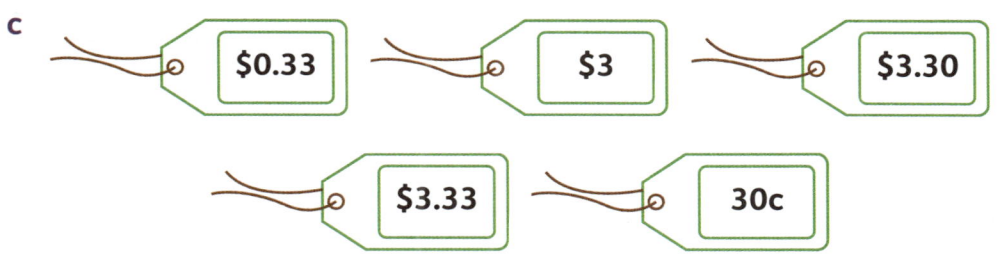

| $0.33 | $3 | $3.30 |

| $3.33 | 30c |

d

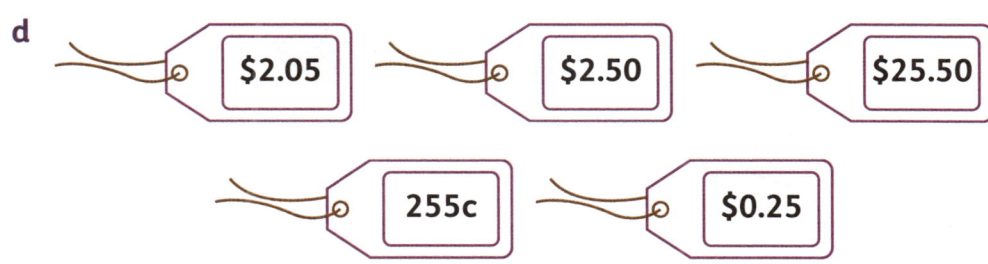

| $2.05 | $2.50 | $25.50 |

| 255c | $0.25 |

💡 **Problem solving**

3 **a** Mika has 150 two-cent coins. How much money is this?

  b Josh has 30 five-cent coins. How much money is this?

  c A bag of sweets costs 85c. How much do two bags of sweets cost?

  d Nico paid 90c for two lollipops. How much did one lollipop cost?

➡ *Workbook page 34*

Totals and change

There are 100 pence (p) in a pound (£1.00).

Pete has £1.00. He wants to buy an apple for 63p. How much change will he get?

$63 + \boxed{} = 100$

He will get 37p change.

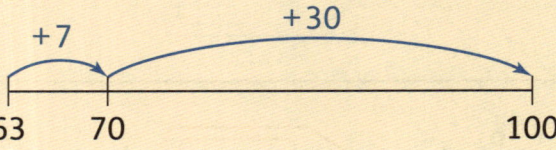

1. How much more does each pupil need to make £1.00?

Nick	Jenna	Patience	Zayed	Mariam
43p	29p	55p	39p	79p
Alexis	Farid	Zuki	Selma	Tom
71p	92p	84p	76p	49p

💡 **Problem solving**

2. How much change will you get from £100 if you buy each item?

a

£38

b

£73

c

£29

3. a Which two items can you buy with £100 and still get change?

   b If you buy the sunglasses and the headphones, how much more money do you need?

4. Imagine that you have £100. What will you buy? Find out the prices of some things you would like to buy. Work out how much you will have left from your imaginary £100.

   If you would like something more expensive than £100, work out how much more you will need.

➡ *Workbook page 35*

<table>
<tr><td></td><td># Mass</td></tr>
</table>

Measure in kilograms

> ### 💭 Think and share
>
> - What do you notice about the markings on these scales?
>
> - What is the same? What is different?
>
> - What kinds of things do we use scales to measure?
>
> - What kinds of scales do people use at home to weigh things?

1 What kind of scale do you have in your classroom or at home? Which of these scales look similar to yours? Say how you think each scale works.

a  b  c  d

2 You will need a measuring scale and five objects to weigh.

Make a table like this one, but with five rows. Estimate the mass of each object. Write your estimate. Then weigh each object and write its actual mass.

> If you are using a balance scale, you will also need 1 kg weights.

Object	Estimate	Actual mass
	kg	kg
	kg	kg

➡ *Workbook page 36*

Kilograms and grams

Each **kilogram** on this scale is divided into smaller units. The smaller units are called **grams**.

1 kg = 1000 g $\quad$ $\frac{1}{2}$ kg = 500 g

On this scale, each small division represents 100 grams.

The fruit on the scale is 1 kg and 500 g.

We can write this as 1 kg 500 g or 1500 g.

1 The units (kilograms or grams) are missing from these objects. Write each mass with the correct unit.

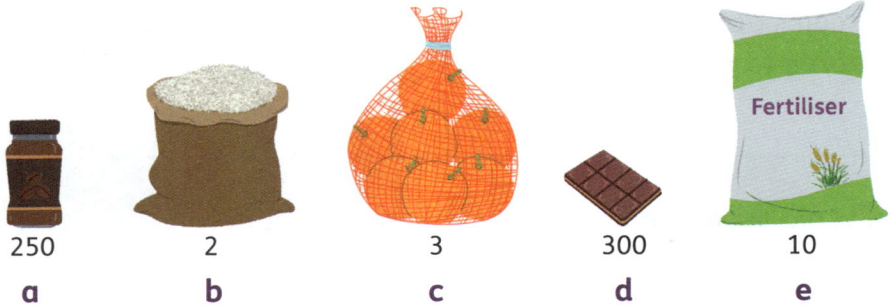

250 $\qquad$ 2 $\qquad$ 3 $\qquad$ 300 $\qquad$ 10

a $\qquad$ **b** $\qquad$ **c** $\qquad$ **d** $\qquad$ **e**

Problem solving

Remember when you added multiples of 100 to make 1000. Remember that 1 kg = 1000 g.

2 Make pairs of items with a total of exactly 1 kg.

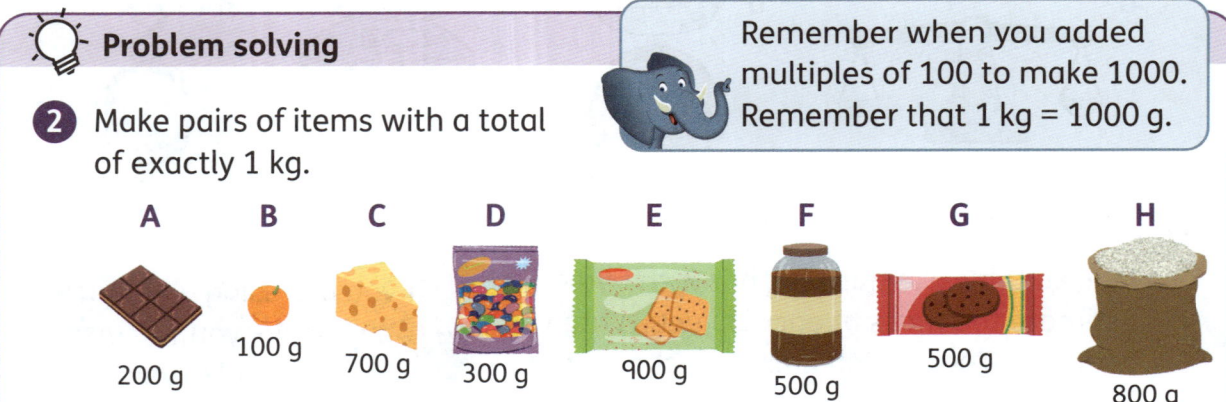

A $\qquad$ B $\qquad$ C $\qquad$ D $\qquad$ E $\qquad$ F $\qquad$ G $\qquad$ H

200 g $\quad$ 100 g $\quad$ 700 g $\quad$ 300 g $\quad$ 900 g $\quad$ 500 g $\quad$ 500 g $\quad$ 800 g

3 Find things at school or at home with these masses:

   a about 100 g $\qquad\qquad$ **b** about 200 g

   c about $\frac{1}{2}$ kg $\qquad\qquad\quad$ **d** about 1 kg

Read scales

1 What is the mass of the items on each scale? Round to the nearest kilogram.

a

b

c

d

e

f

2 Work out the total mass if you buy:

   a the bananas, potatoes and carrots

   b the apples, pears and cabbages

   c everything

   d everything except the carrots.

3 Choose four of the scales.

   a What is the total mass of the items on the four scales?

   b What is the total mass of two lots of each item?

➡ Workbook page 37 and page 38

Multiplication and division

Revise multiplication and division

Think and share

How many eggs can you see?

What do you notice?

What do you wonder?

Make up some questions about the numbers you can make.

Each flower has 5 petals.

Here are the multiplication and division facts for this picture.

$5 \times 3 = 15$
$3 \times 5 = 15$
$15 \div 3 = 5$
$15 \div 5 = 3$

You can also show these facts as **arrays**.

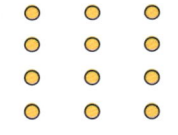

1 Write the multiplication and division facts for each picture.

a

Each seedling has two leaves.

b

Each pod has five peas.

c

Each vase has three flowers.

d

Each fork has four prongs.

e

Each pot has two seedlings.

f

Each row has four cabbages.

➡ *Workbook page 39*

Times tables

● ● 2 ● ● 4 ● ● 6 ● ● 8	○ ○ ○ ○ ○ 5 ○ ○ ○ ○ ○ 10 ○ ○ ○ ○ ○ 15 ○ ○ ○ ○ ○ 20	● ● ● ● ● ● ● ● ● ● 10 ● ● ● ● ● ● ● ● ● ● 20 ● ● ● ● ● ● ● ● ● ● 30 ● ● ● ● ● ● ● ● ● ● 40
These dots are arranged in rows of 2.	These dots are arranged in rows of 5.	These dots are arranged in rows of 10.

1 Copy and complete.

     a $2 \times 1 = \Box$ $1 \times 2 = \Box$ **b** $2 \times 3 = \Box$ $3 \times 2 = \Box$

     c $2 \times 4 = \Box$ $4 \times 2 = \Box$ **d** $2 \times 5 = \Box$ $5 \times 2 = \Box$

2 Write the answers.

     a double 6   **b** double 8   **c** double 10

     d double 9   **e** double 7   **f** double 5

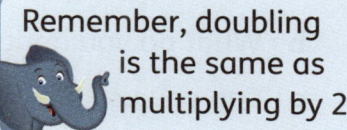

Remember, doubling is the same as multiplying by 2.

3 Multiply each number by 5.

     a 0   **b** 3   **c** 6   **d** 8   **e** 10   **f** 9

4 Write the 5 times table. Ask a partner to check your work.

5 Copy and complete.

     a $10 \times 6 = \Box$ **b** $10 \times 3 = \Box$ **c** $10 \times 8 = \Box$

     d $10 \times 10 = \Box$ **e** $10 \times \Box = 50$ **f** $\Box \times 4 = 40$

6 When you multiply by a number, your answer is a multiple.

     a Look at your answers to questions 1 and 2.
         What do you notice about multiples of 2?

     b Look at your answers to question 4.
         What do you notice about multiples of 5?

     c Look at your answers to question 5.
         What do you notice about multiples of 10?

➡ Workbook page 40

More times tables

Nisha has arranged number frames to show groups of 3.

1 group of 3 = 3 $1 \times 3 = 3$

2 groups of 3 = 6 $2 \times 3 = 6$

1 Write the 3 times table, from 1×3 up to 12×3. Use the number frames to help you.

2 **a** Complete the first four multiples in the 6 times table. Use the number frames to help you.

$6 \times 1 = \square$ $6 \times 2 = \square$ $6 \times 3 = \square$ $6 \times 4 = \square$

b What do you notice about the 3 times table and the 6 times table?

c Can you use the 3 times table to help you work out 6×5?

3 Use the number frames above to help you work out these divisions.

a $24 \div 3 = \square$ **b** $24 \div 4 = \square$

c $24 \div 6 = \square$ **d** $24 \div 2 = \square$

4 Look at these number frames. What patterns do you notice?

5 **a** Write multiplication and division facts about the 4 times table. Use the number frames to help you.

b Write the first ten multiples of 8.

c What do you notice about multiples of 4 and multiples of 8?

➡ *Workbook page 41*

Revise division into groups

Equal sharing is called division. **Divide** these bottle tops into groups of two.

We can make 6 equal groups of 2.

$12 \div 2 = 6$

Layla uses bottle tops to make different craft items. She uses:

5 bottle tops for
a flower

2 bottle tops for
a pair of earrings

6 bottle tops for a
family of monsters.

1 How many flowers can she make from:

  a 25 bottle tops   **b** 45 bottle tops   **c** 50 bottle tops?

2 How many pairs of earrings can she make from:

  a 10 bottle tops   **b** 12 bottle tops   **c** 18 bottle tops?

3 How many monster families can she make from:

  a 12 bottle tops   **b** 18 bottle tops   **c** 30 bottle tops?

4 Collect some bottle tops and make your own craft items. Explain how many bottle tops you used. Use division to work out how many items you could make with 100 bottle tops.

Multiplication and division facts

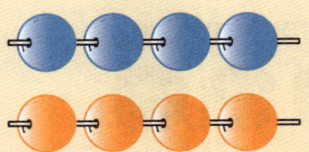 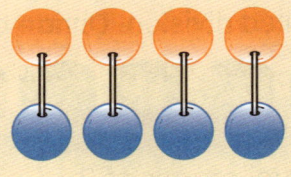

2 groups of 4 = 8
2 × 4 = 8

4 groups of 2 = 8
4 × 2 = 8

We can divide 8 into:

2 equal groups
OR
groups of 2

$8 \div 2 = 4$

4 equal groups
OR
groups of 4

$8 \div 4 = 2$

1 Use these pictures to help you write sets of multiplication and division facts.

a

b

c

d

Division facts

Division is the inverse of multiplication, just like addition is the inverse of subtraction. You can use **inverse operations** to help you solve problems.

1	2	3	4	5	6	7	8	9	10
11	12	13	14	15	16	17	18	19	20
21	22	23	24	25	26	27	28	29	30
31	32	33	34	35	36	37	38	39	40
41	42	43	44	45	46	47	48	49	50
51	52	53	54	55	56	57	58	59	60
61	62	63	64	65	66	67	68	69	70
71	72	73	74	75	76	77	78	79	80
81	82	83	84	85	86	87	88	89	90
91	92	93	94	95	96	97	98	99	100

$16 \div 8 = \square$

What times 8 makes 16?
$2 \times 8 = 16$ so $16 \div 8 = 2$

$30 \div 6 = \square$

What times 6 makes 30?
$5 \times 6 = 30$ so $30 \div 6 = 5$

1 Use inverse operations to help you work these out.

a $3 \times 10 = 30$

$30 \div 10 = \square$

$30 \div 3 = \square$

b $8 \times 3 = 24$

$24 \div 8 = \square$

$24 \div 3 = \square$

c $5 \times 9 = 45$

$45 \div 5 = \square$

$45 \div 9 = \square$

d $4 \times 7 = 28$

$28 \div 4 = \square$

$28 \div 7 = \square$

e $5 \times 10 = 50$

$50 \div 5 = \square$

$50 \div 10 = \square$

f $3 \times 7 = 21$

$21 \div 3 = \square$

$21 \div 7 = \square$

2 Complete these divisions. Use times tables facts you already know to help you.

a $15 \div 3 = \square$

b $12 \div 3 = \square$

c $14 \div 2 = \square$

d $18 \div 6 = \square$

e $24 \div 8 = \square$

f $20 \div 4 = \square$

g $32 \div 4 = \square$

h $25 \div 5 = \square$

i $90 \div 10 = \square$

➡ *Workbook page 42*

Factors

Do you remember sharing 12 bottle tops into equal groups? The bottle tops were already in:

1 group of 12   OR   12 groups of 1

You can also arrange the bottle tops in:

4 groups of 3   OR   3 groups of 4   2 groups of 6   OR   6 groups of 2

1, 2, 3, 4, 6 and 12 are **factors** of 12. This means that we can share a group of 12 equally into groups of each number, with none left over.

1 Use these pictures to help you work out the factors of:

a

18

b

24

2 Use collections of objects to work out all the factors of:

a 63   **b** 75   **c** 100

Use objects or sketches to help you explain.

💡 **Problem solving**

3 True or false? Explain your answers.

a An odd number can have 2 as a factor.

b All whole numbers have 1 as a factor.

c A number can have a whole number factor greater than itself.

➡ *Workbook page 43*

Multiples

$5 \times 2 = 10$

10 is a **multiple** of 2. 10 is also a multiple of 5.

So, when we count in groups of 2 or in groups of 5, one of the numbers we will count is 10.

2	4	6	8	10	12	14	16	18	20

10 is the fifth multiple of 2. In other words, 5 groups of 2 make 10.

5	10	15	20	25	30	35	40	45	50

10 is the second multiple of 5. In other words, 2 groups of 5 make 10.

1 a Count in multiples of 4.

4	8	12	16	20	24	28	32	36	40
44	48	52	56	60	64	68	72	76	80

   b What pattern do you notice in the ones place?

   c Can you predict the numbers in the next row of the track?

   d Are all even numbers multiples of 4? How do you know?

2 a Draw your own track with the first twenty multiples of 5.

   b Use your track to help you say the 5 times table.

   c What patterns do you notice that can help you multiply by 5?

   d Do you think the number 785 is a multiple of 5? How do you know?

3 a Count in multiples of 50.

50	100	150	200	250	300	350	400	450	500
550	600	650	700	750	800	850	900	950	1000

   b What patterns do you notice in the sequence of multiples of 50?

   c How can you use this track to help you count in multiples of 100?

   d Look at the number 650. Is it a multiple of 50? Is it a multiple of 100? Explain how you can work this out.

💡 Problem solving

Test this statement with some factors and multiples that you know.

4 True or false? If  is a multiple of ,

that means  is a factor of .

▶ *Workbook page 44 and page 45*

Divisibility

We have seen that the factors of 10 are 1, 2, 5 and 10.

So, 10 is **divisible** by 1, 2, 5 and 10. We can divide 10 equally into groups of these sizes, without any **remainder**.

10 is not divisible by 3

We can use tests of divisibility to work out whether numbers are divisible by a given number. For example: if a number ends in 0, it is divisible by 10.

These numbers are all divisible by 10: | 50 | 1000 | 1280 | 470 | 350 |

1. a Why does a zero at the end of a number tell us that a number is divisible by 10?

   b If a number ends in 00 (for example, 500, 900, 200), what number is it divisible by? How do you know?

2. The first eight multiples of 25 are: 25, 50, 75, 100, 125, 150, 175, 200.

   a Write the next eight multiples of 25, starting with 225.

   b What pattern do you notice in multiples of 25?

3. How do you know whether a number is divisible by 5?

4. Which statement is correct and which is incorrect? How can you check?

   > Any number that is divisible by 10 is also divisible by 5.

   > Any number that is divisible by 25 is also divisible by 100.

Problem solving

5. Write a number that is:

   a divisible by 3 but not by 5

   b divisible by 4 but not by 3

   c divisible by 5 but not by 10

   d divisible by 5 but not by 25.

➡ *Workbook page 46 and page 47*

Multiples of 10

What is 9×10?
Think of this as 9 tens. Write the 9 in the tens place.
Write a 0 in the ones place as a placeholder.

Tens	Ones
	9
9	0

$9 \times 10 = 90$

What is 18×10? Think of this as 18 tens.

18 tens is the same as 1 hundred and 8 tens.

Hundreds	Tens	Ones
	1	8
1	8	0

$18 \times 10 = 180$

When we multiply a number by 10, the digits move one place left.
We write 0 as a placeholder in the ones place.

1 Try to do these multiplications in your head.

a 7×10 b 16×10 c 19×10 d 20×10

e 22×10 f 25×10 g 34×10 h 36×10

i 45×10 j 52×10 k 59×10 l 61×10

2 Crayons are sold in boxes of 10. How many crayons are there are in:

a 12 boxes   b 21 boxes   c 39 boxes?

3 A school ordered some boxes of 10 crayons. They received 420 crayons. How many boxes did they order?

Problem solving

4 Each shape represents a number. You multiply the shape at the top of the column by the shape at the beginning of the row to make the number.

Can you work out the value of each shape?

➡ *Workbook page 48 and page 49*

Doubling and halving

What is double 15?
double 10 = 10 × 2 = 20
double 5 = 5 × 2 = 10
20 + 10 = 30

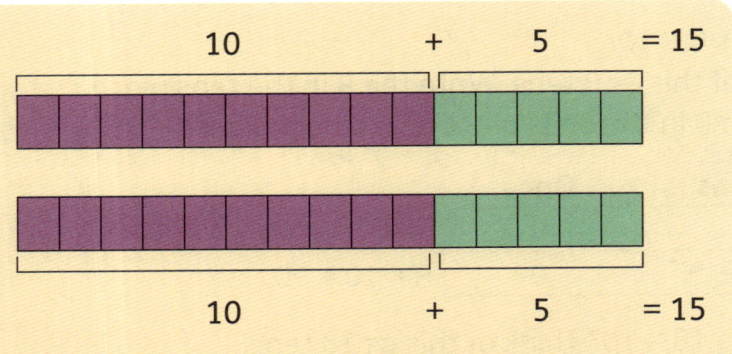

10 + 5 = 15

Halving is the inverse of doubling.
What is half of 18?
$\frac{1}{2}$ of 10 = 10 ÷ 2 = 5
$\frac{1}{2}$ of 8 = 8 ÷ 2 = 4
5 + 4 = 9

10 + 5 = 15

1 Double each number. Estimate before you work it out.

   a 11   **b** 12   **c** 13   **d** 14

   e 15   **f** 16   **g** 17   **h** 18

You could draw a bar model to help you double.

2 Copy and complete these statements.

   a ☐ is half of 20   **b** ☐ is half of 50

   c ☐ is half of 100   **d** ☐ is half of 70

3 What is half of each amount?

   a  £12 **b**  £16 **c**  £18

4 I need 13 metres of fabric to make one set of curtains.
Estimate, then work out how much fabric I need to make:

   a two sets of curtains   **b** four sets of curtains.

5 Mr Jonas has 28 metres of rope. He wants to cut the rope in half.
Estimate, then calculate how long each piece will be.

6 On a 32 km cycle, Salman stops to rest after the first 12 km.

   a Has he cycled half the distance when he stops?

   b How much further does he need to cycle to be halfway?

➡ *Workbook page 50*

Perimeter and area

Perimeter

💭 Think and share

The **perimeter** is the distance around the outside of a 2D shape.

Lindy drew these shapes.

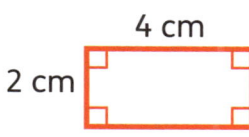

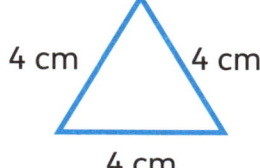

Just by looking, estimate the answer to this question:

• Does one shape have a longer perimeter than the other shape?

We can work out the perimeter by adding up the lengths of all the sides.

How do you know the lengths of the unlabelled sides of the rectangle? What do you know about rectangles that can help you?

With a partner, work out the perimeter of each shape. Was your estimate correct?

1 Work out the perimeter of each shape. Add up the lengths of all the sides.

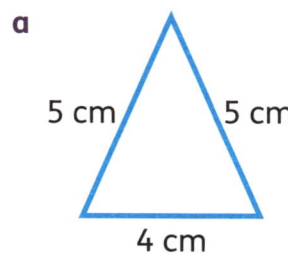

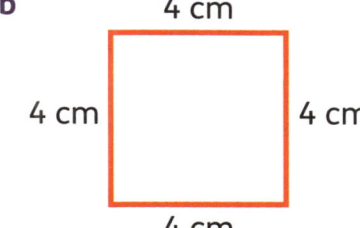

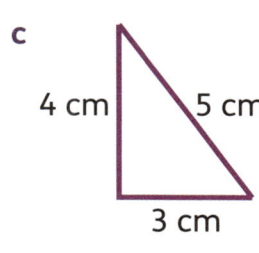

2 Each grid square represents a square that is 1 cm long and 1 cm wide. Work out the perimeter of each shape drawn on the grid.

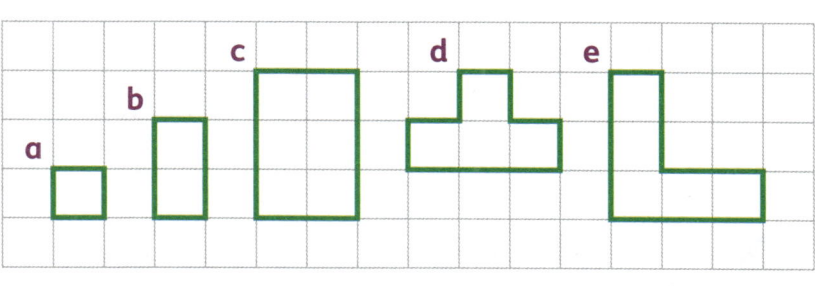

➡️ *Workbook page 51 and page 52*

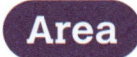

Area

> The **area** is the amount of space that a 2D shape takes up.
>
> We measure area in **square units**.
>
> You can count squares to work out the area of a 2D shape.

1 How many squares does each rectangle take up? Write the areas. Write the letters in order, from the rectangle that takes up the most squares to the rectangle that takes up the fewest squares.

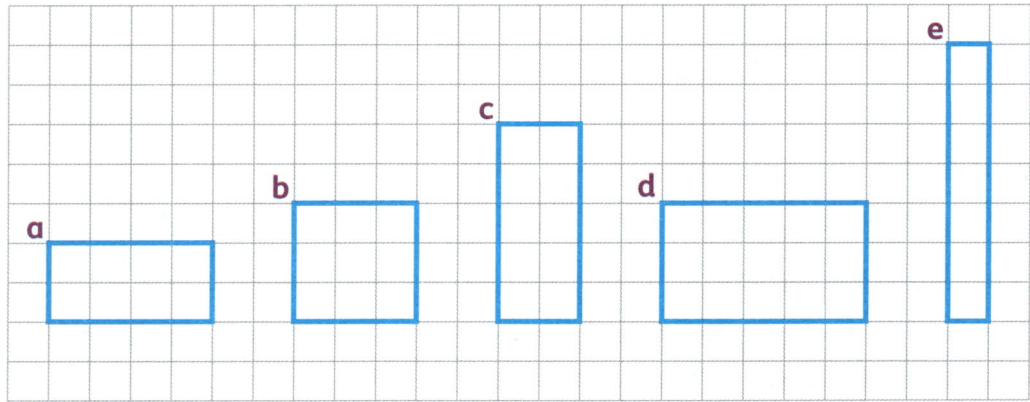

2 Look at the rectangles in question 1. Each square on this grid represents a square 1 cm long and 1 cm wide.

   a In pairs, work out the length of each side of the rectangles.

   b Look at the side lengths and the number of squares that each rectangle takes up. What pattern do you notice?

3 Count the squares to find the areas of these shapes.

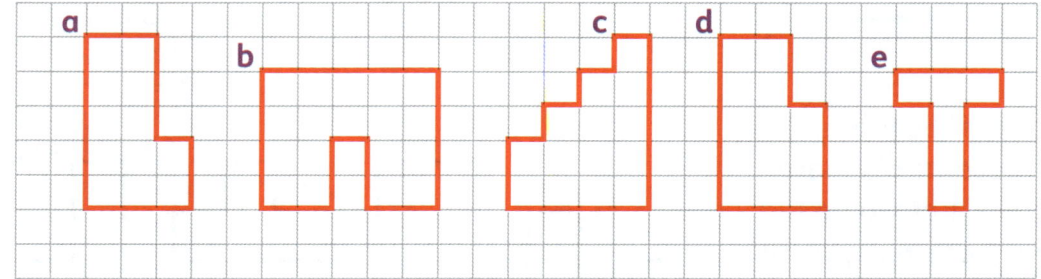

4 Look at your answer to question 2b. Do you notice the same pattern with the shapes in question 3? What is similar? What is different?

Work with perimeter and area

We measure area in square units.

☐ = 1 square centimetre

We write 1 square centimetre like this: 1 cm².

This shape takes up 6 square centimetres or 6 cm².

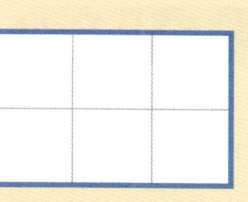

Each grid square in questions 1 and 2 represents 1 cm².

1 Find the areas of these shapes. Give your answers in cm².

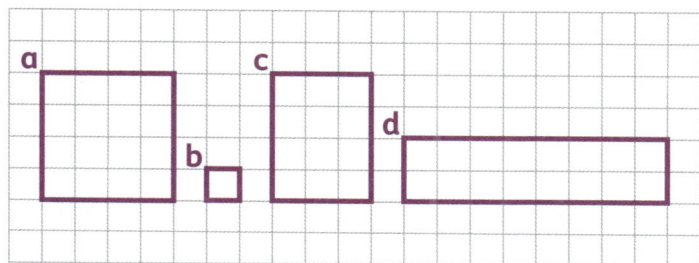

2 Which of these shapes do you think has the largest area? Estimate, then count the squares. Discuss what you notice with a partner.

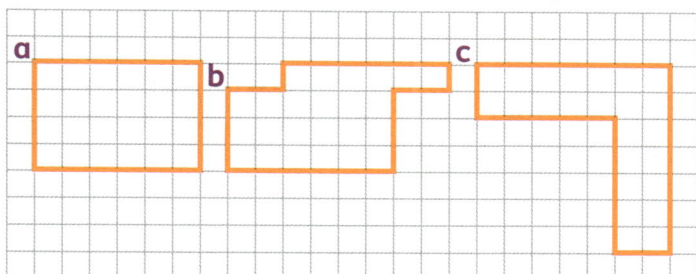

Problem solving

3 Is it possible to make shapes that have the same perimeter but different areas? Is it possible to make shapes that have the same area but different perimeters?
Draw sketches on squared paper to investigate.

4 An ant crawls up these stairs from point A to point B. Work out the total distance the ant crawls.

B

4 metres

A   8 metres

➡ *Workbook page 53 and page 54*

Data

Charts and tables

Think and share

CLASS GROUPS POINTS CHART

WEATHER CHART						
Monday	Tuesday	Wednesday	Thursday	Friday	Saturday	Sunday

CHORES CHART	
Dishes	⊞⊞ ⊞⊞ II
Laundry	III
Floor	⊞⊞ I

- Where have you seen charts or tables like these?
- What would help you to understand these charts better?

1. In the chores chart, how many times did the family:

   a wash the dishes   b do the laundry

   c clean the floor?

   d After 25 chores, the family will order pizza. How many more chores must they complete before they order pizza?

   > I = 1 ⊞⊞ = 5. We group **tally marks** in groups of 5 to help us count the marks easily.

2. Draw these numbers as tally marks.

   a 2   b 5   c 10   d 17   e 20

3. Find out the favourite vegetables in your class. Use a **tally table** like this.

Favourite vegetable	Number of people
Spinach	
Carrots	
Beans	
None of the above	

➡ *Workbook page 55*

Read tables

Some children wrote how much television they watched each day for a week. They wrote the information in this table.

Name	Hours of TV watched						
	Sun	Mon	Tue	Wed	Thur	Fri	Sat
Sarah	0	2	2	0	1	2	2
Ranjit	1	1	3	1	2	3	0
Mark	2	2	2	3	3	4	6
Tara	3	1	3	1	1	0	3

1 Write answers to these questions.

   a For how long did Tara watch TV on Friday?

   b For how long did Mark watch TV on Saturday?

   c For how long did Ranjit watch TV on Wednesday?

   d For how long did Tara watch TV on Sunday?

   e Who watched the most TV during this week?

   f On which days did Sarah watch no TV?

   g Who did not watch TV on Saturday?

   h On which day did the group watch the most TV?

2 You are going to make your own table to record how long you watch TV or listen to music each day.

- Choose watching TV or listening to music.

- Estimate before you start.

- Every day for a week, record how long you watch TV or listen to music.

   a How close was your estimate?

   b On which days do you watch TV or listen for the longest time?

   c On which days don't you watch TV or listen to music?

   d Compare your table with your classmates' tables. Do you spend more or less time watching TV or listening to music than your classmates?

Pictograms

Rina has an ice cream shop. She drew a **pictogram** to show which ice cream flavours she sold in her shop.

Look at the key. Each 🍦 picture means 2 ice creams.

Half a picture 🍦 means half of 2, which is 1.

Ice cream cones sold	
Vanilla	🍦🍦
Chocolate	🍦🍦🍦🍦
Strawberry	🍦🍦
Neapolitan	🍦
Banana	🍦🍦🍦
Mint choc chip	🍦🍦🍦

Key: 🍦 = 2 cones

1 How many of these ice creams did Rina sell?

   a vanilla   **b** banana   **c** mint choc chip   **d** Neapolitan

2 **a** Which ice cream flavour was the most popular?

   b Which ice cream flavour was the least popular?

3 How many ice creams does each picture show?

   a    **b**    **c**

4 Draw your own pictures to show sales of:

   a 4 ice creams   **b** 10 ice creams   **c** 13 ice creams **d** 21 ice creams.

💡 Problem solving

5 Kofi kept a record of how many books he read in four months. He drew this pictogram.

   a How many books did Kofi read altogether?

   b In which two months did he read a total of 10 books?

   c Kofi wants to read 24 books in total by the end of September. How many books must he read in August and September?

Month	Books read
April	📕📕📕
May	📕📕📕📕📕
June	📕📕📕📕
July	📕📕📕📕📕📕

Key: 📕 means 1 book

➡ *Workbook page 56 and page 57*

Read a bar chart

1. This **bar chart** shows the favourite sports of pupils in a class.

   Write answers to these questions.

   a Which is the most popular sport?

   b How many pupils like swimming best?

   c How many pupils like gymnastics best?

   d What is the third most popular sport?

   e How many pupils chose basketball?

   f How many more pupils like football than netball?

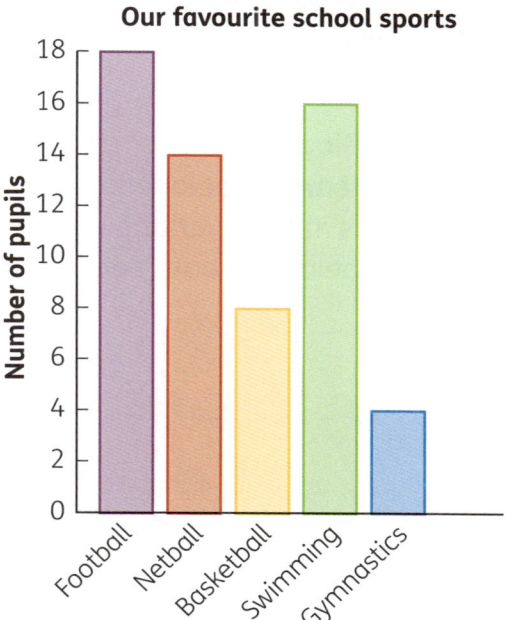

Our favourite school sports

2. Amy asked her classmates which fillings they like best in their sandwiches. She drew tally marks for her results, but she did not complete the table.

Cheese	ЖЖ II	7
Egg	IIII	
Chicken	ЖЖ I	
Vegetables	II	
Jam	ЖЖ ЖЖ II	

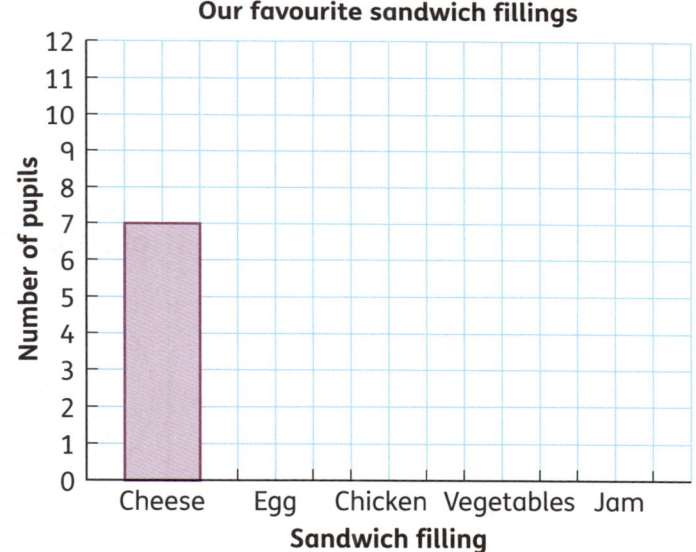

Our favourite sandwich fillings

   a Copy Amy's tally table and fill in the missing totals.

   b Copy Amy's bar chart and complete it for her.

3. Find out which sandwich fillings your classmates like best.

   a Make a tally table to record the information.

   b Draw a bar chart to show the information.

➡ *Workbook pages 58 to 61*

Venn diagrams

A **Venn diagram** uses circles to show which **elements** belong in a set. Elements in the overlapping section belong in both sets.

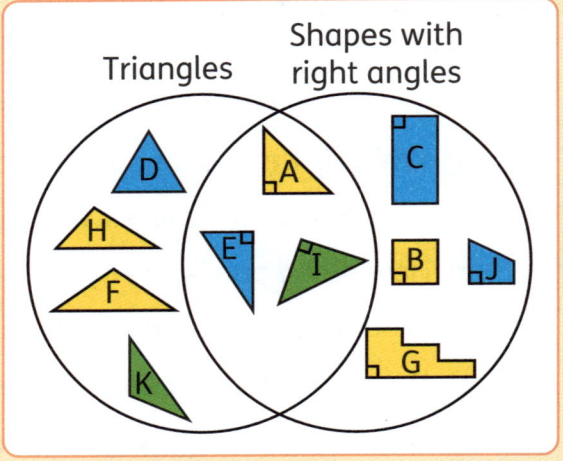

1. Look at the Venn diagram. Write the letters of shapes that:

   a belong in the set of triangles

   b belong in the set of right-angled shapes

   c belong in both sets.

2. Where does this shape fit in the Venn diagram?

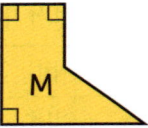

3. Look at this Venn diagram.

   a List all the elements of Set A.

   b List all the elements of Set B.

   c A number is missing from the part where Set A and Set B overlap. What is it?

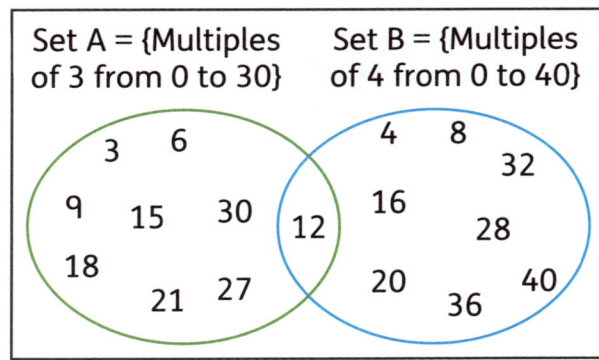

   d Dan says: '36 is a multiple of 3 and a multiple of 4, so it should be an element of both sets.' What is his mistake?

4. Draw your own Venn diagram to show these sets:

   Set M = {Even numbers from 0 to 20}

   Set Q = {Multiples of 5 up to 25}

➤ *Workbook page 62*

Carroll diagrams

A **Carroll diagram** helps us to sort information. Each box in the table matches two of the rules. Can you work out the rules for each box?

	I have long hair	I do not have long hair
I have brown hair	Indira	Melissa
I do not have brown hair	Alex	Leroy

1 Jacobus asked his friends whether they like oranges and apples.

	I like apples 🍎😊	I do not like apples 🍎☹️
I like oranges 🍊😊	Sandy Koos Lucy Peter	Ahmed Bali
I do not like oranges 🍊☹️	Mandla Dan Connie	Teresa

a How many of Jacobus's friends like apples and oranges?

b Who doesn't like apples or oranges?

c Who likes apples, but not oranges?

d Who likes oranges, but not apples?

2 Ask at least six classmates whether they like carrots and peas.
Make a Carroll diagram to record their answers.

3 Copy this Carroll diagram.

	I like basketball 🏀😊	I do not like basketball 🏀☹️
I like swimming 〰️😊		
I do not like swimming 〰️☹️		

a Ask ten classmates whether they like basketball and swimming.

b Make tally marks in your table to show your classmates' answers.

➤ Workbook page 63 and page 64

Mixed practice 2

1 **a** Write three different additions that each make a total of 10.

   b Use your answers to part **a** to help you write three additions that make a total of 100.

2 **a** Noah has $589. He spends $125. How much does he have left?

   b Kirby has $200. She earns another $500, and then $200 more. How much more does she need to make $1000? Show your working.

3 Cards are sold in packs of 10. I need 244 cards.

   a How many packs should I buy?

   b Why should I round up to the nearest 10, not down?

4 Complete these additions and subtractions. You can use any strategy, but show your working.

   a 53 + 60   **b** 190 + 40   **c** 400 + 150

   d 189 – 45   **e** 748 – 99   **f** 358 – 197

5 For each amount, how much more do I need to make $2?

   a $1.50   **b** $1.90

   c $1.25   **d** 20c

 There are 100 cents in $1.

6 Estimate whether the mass of each item is less than 1 kg, more than 1 kg or about 1 kg.

   a    **b**    **c**

   a bag of potatoes   a cabbage   a pair of scissors

7 An apple has a mass of about 100 g. About how many apples make 1 kg?

8 Read the mass shown on each scale.

   a    **b**    **c**

9 Write the multiplication and division facts you can make from this array.

10 a Write the first ten multiples of 3.

   b Write the first ten multiples of 5.

   c 947 is not a multiple of 5. How can you tell?

11 James draws a square. Each side is 5 cm long. James says the perimeter is 25 cm². What is his mistake?

12 Work out the area of each shape in squares.

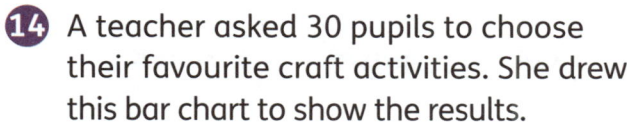

13 Jasmyn has a cube-shaped box. She places it on a piece of 1 cm-squared paper like this.

What area of the squared paper does the cube cover?

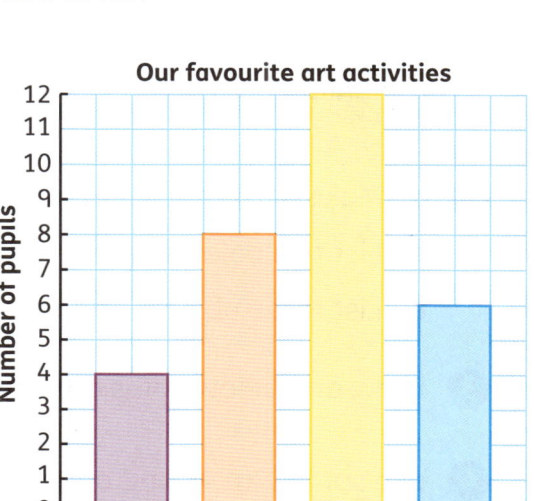

14 A teacher asked 30 pupils to choose their favourite craft activities. She drew this bar chart to show the results.

   a Which activity did most pupils choose?

   b How many pupils chose clay?

   c How many more chose painting than drawing?

Our favourite art activities

Number of pupils vs *Art activities* (Clay, Painting, Junk, Drawing)

15 Draw a Venn diagram to show these sets.

A = {Even numbers from 99 to 121}

B = {Multiples of 5 from 89 to 127}

16 Copy this table. Write something that fits in each block.

	Edible	Not edible
Red		
Not red		

Edible means you can eat it.

3D shapes

Identify 3D shapes

> **Think and share**
>
> - What do you see?
> - Where have you seen these 3D shapes before?
> - What things do you use that have these shapes?
> - Which shapes can you roll?
> - What food packets and boxes have you seen that are these shapes?

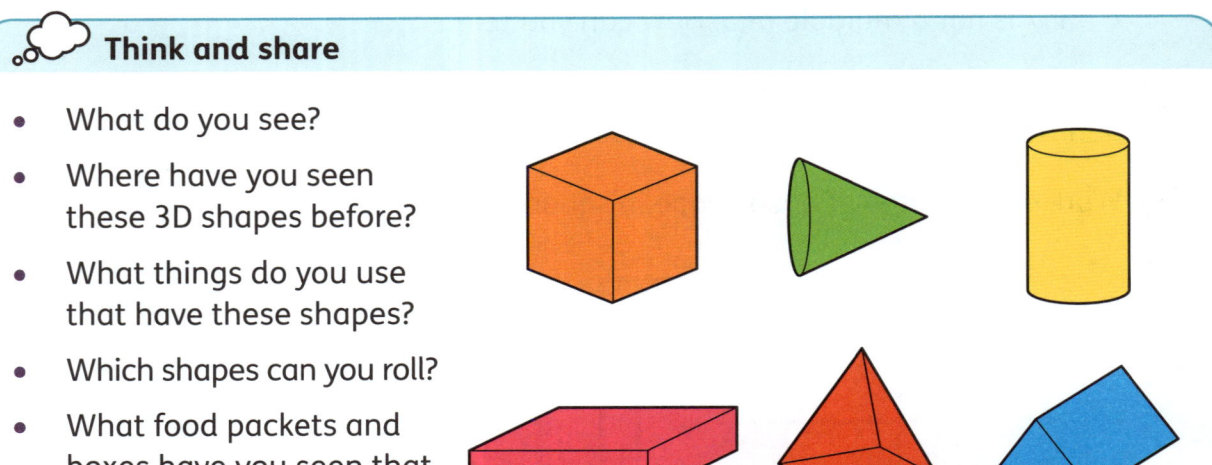

1 Do you know all these words? Match each word to one of the shapes above.

cuboid   cone   cylinder   cube   prism   pyramid

2 **a** Make a list of objects at home or school that are cuboids.

b List five objects at the supermarket that are cylinders.

c List three objects at home or at school that are cubes.

3 Draw an object that is a **triangular prism** shape.

4 Use some straws and adhesive tack or tape to make models of the shapes you can see on this page.

straws   adhesive tack   tape

Faces, edges and vertices

On a 3D shape:

- the flat surfaces are called **faces**
- the places where two faces meet are called **edges**
- the corners where the edges meet are called **vertices**.

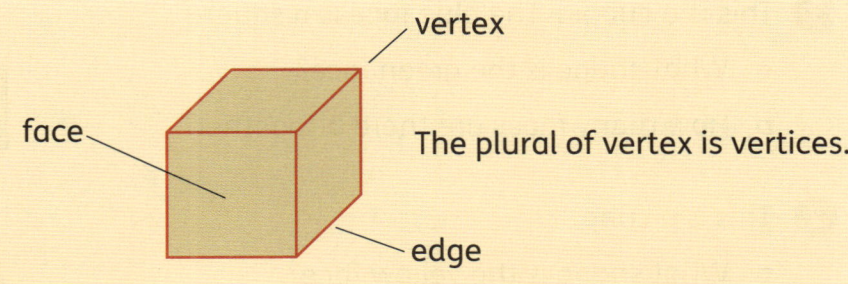

vertex

face

edge

The plural of vertex is vertices.

In this drawing of a **cube**, you cannot see all the faces, edges and vertices. Some are hidden from view.

1 Look at a real cube. How many faces, edges and vertices are hidden in the drawing of a cube above?

2 a How many faces, edges and vertices can you see in each shape below?

b Discuss with a partner. How many faces, edges and vertices are hidden on each shape?

A 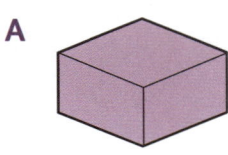   B 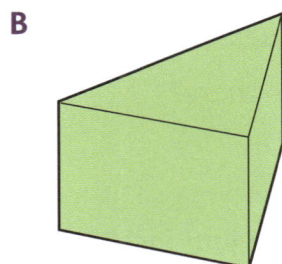   C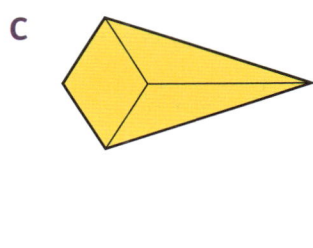

💡 Problem solving

3 What shape am I? Identify the shape that matches each description.

a I have no vertices. I have two faces that are circles. I can roll.

b I have six faces. They are all square.

c I have five faces. One face is a square and the other faces are triangles.

d I have six faces and eight vertices. Not all my faces are square.

➡ *Workbook page 65*

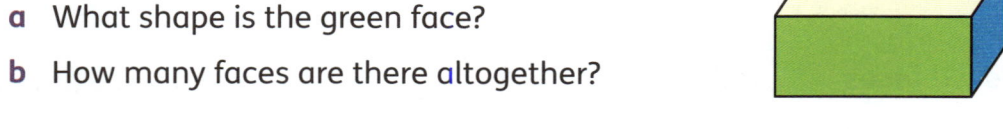

Explore 3D shapes

1 This is a cuboid. The blue face is a square.

   a What shape is the green face?

   b How many faces are there altogether?

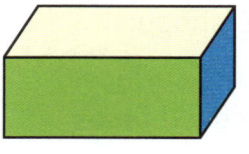

2 This is a cone.

   a What shape is the yellow face?

   b How many vertices does a cone have?

3 This is a cube.

   a What shape is the blue face?

   b What shape is the green face?

   c How many faces does a cube have?

   d How many vertices does a cube have?

4 This is a **square-based pyramid**.

   a What shape is the yellow face?

   b What shape is the red face?

   c How many faces are there altogether?

   d How many vertices are there?

5 This is a cylinder.

   a What shape are the two end faces?

   b How many vertices does a cylinder have?

6 This is a triangular prism.

   a What shape is the red face?

   b How many faces are shaped like this?

   c What shape is the yellow face?

   d How many faces are shaped like this?

   e How many edges are there on the prism?

➡ *Workbook page 66 and page 67*

3D shapes in different positions

Many 3D shapes look different when they are in different positions.

Look at this tin can. How does it look different in each picture? What shapes can you see?

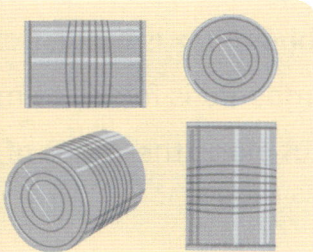

1 Look at each set of 3D shapes. Tell a partner which shape doesn't fit in the set. Explain why.

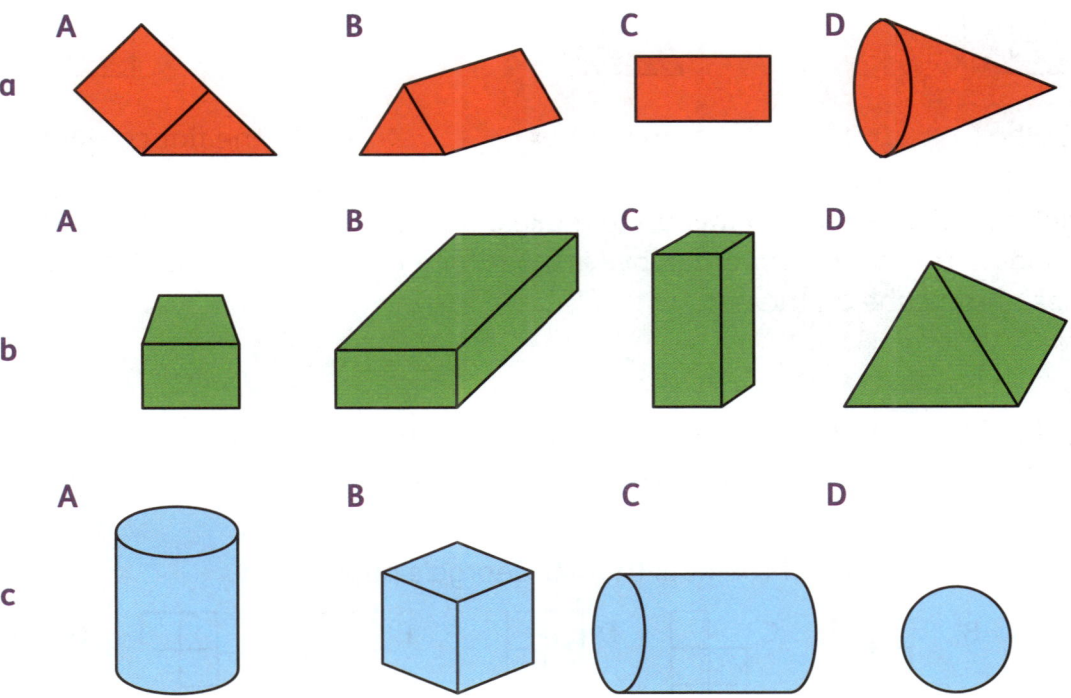

2 Your teacher will give you some containers – cuboid, prism or cylinder shapes.

With a partner, look at each container from different directions. Draw pictures to show it in three different positions.

Crispy rice

Crunchy crisps

Nutty chocolate

Nets of cubes

A cardboard box is made from a flat piece of cardboard.

The cardboard is folded to make a 3D box.

If we cut along the edges of a box and flatten the cardboard, we can see how the box was made.

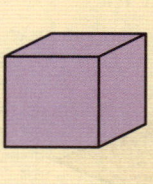

a 3D box

opening the box

the flat cardboard

A flat shape that you can fold up to make a 3D shape is called the **net** of the 3D shape. The flat cardboard above is a net for a cube. You can see the six square faces of the cube on the net.

1 Mr George asked his class to cut open a cube-shaped box and to draw its net. Some pupils didn't cut the box. They just drew what they thought the net would look like.

These are some of the nets that Mr George's pupils drew.

A 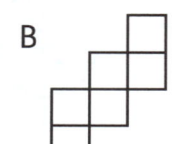 B C 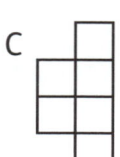 D 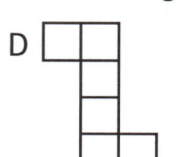 E 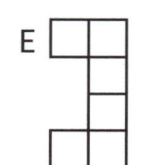 F 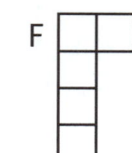 G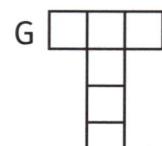

   a Which of these are nets of a cube?

   b How did you decide?

   c How can you tell that net F is not the net of a cube?

2 There are 11 different nets of a cube. Try to find one net that isn't shown on this page. Draw the net on squared paper.

➡ *Workbook page 68*

Position, direction and movement

Position and direction

Positions tell us where things are. We use words like above, below, left, right, next to, between to describe positions.

> **Think and share**
>
> - What is on the top shelf?
> - What is on the bottom shelf?
> - What is to the right of the corn flakes?
>
> Ask your partner questions about the boxes on the shelf. Use position words in your questions.
>
>
>
> Directions tell us which way to go.
>
> We use words like up, down, forwards, backwards, straight on to give people directions.
>
> On a map, we use the **cardinal points** – north, south, west and east – to describe directions.
>
> When do you give or follow directions?

1 Your teacher will show you a world map or globe. Find your country. Identify the nearby countries:

   a to the north   **c** to the east

   b to the south   **d** to the west.

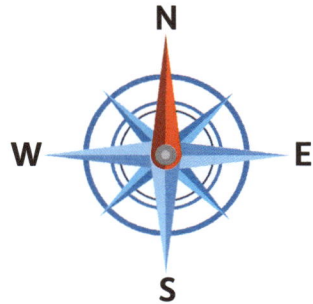

2 **a** Which continent is furthest south on Earth?

   b Which continent is south of Asia and east of Africa?

   c Can we say which continent is furthest to the east or west? Why or why not?

Clockwise and anti-clockwise

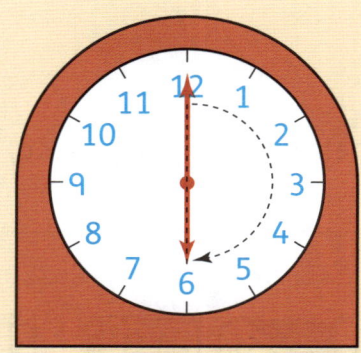

The hands of a clock always turn in the same direction. They pass through the numbers on the clock face in order from 1 to 12.

This direction is called **clockwise**.

The opposite direction is called **anti-clockwise**.

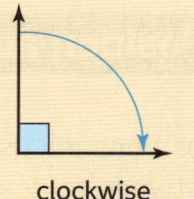

clockwise

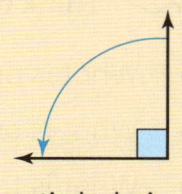

anti-clockwise

1. Play this game in a group of five. One person is the caller. They call out directions using these words.

clockwise	anti-clockwise	quarter turn
half turn	three-quarter turn	full turn

When the caller gives a direction, the group must follow the direction. Anyone who does not follow the direction correctly must sit down. The winner is the last person standing.

Here are some examples of directions.

- Walk clockwise in a circle.
- Walk anti-clockwise in a circle.
- Make a clockwise movement with your right hand.
- Make a quarter-turn to the right.
- Make a full turn, then take one step forward.
- Make a half turn, then take two steps to the left.
- Make anti-clockwise circles with your head.

Make up any other directions that you wish.

➡ *Workbook page 69 and page 70*

Position on a grid

In this classroom, each pupil has a locker.

The columns are labelled A to D.

The rows are labelled 1, 2 and 3.

Each locker has its own name made from a letter and a number.

To find the name, we go across to a letter and up to a number.

The green book is in locker A3.

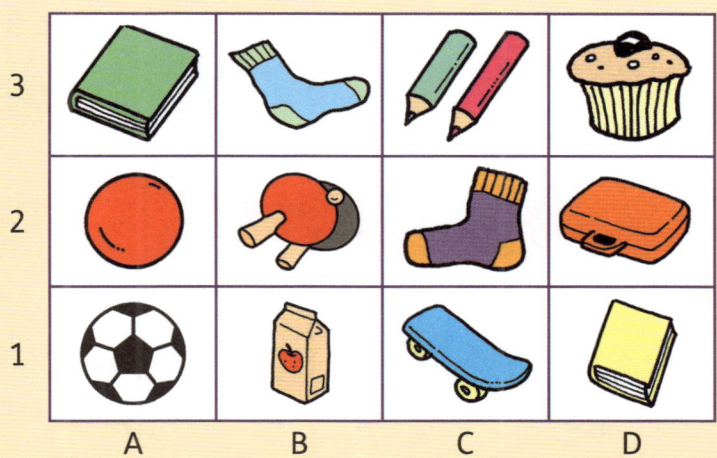

1 Look at the lockers above.

   a What is in B1?

   b What is in C3?

   c In which locker is the skateboard?

   d In which locker is the cake?

2 Write the position of:

   a the rectangle

   b the cone

   c the arrow

   d the pentagon

   e the cube

   f the cuboid.

3 What is the name of the shape in each of these positions?

   a B3   **b** D2   **c** A2

   d C4   **e** B1   **f** D3

➡ *Workbook page 71 and page 72*

Move a shape

This shape has moved on the grid. It has moved 3 dots to the right and 4 dots down.

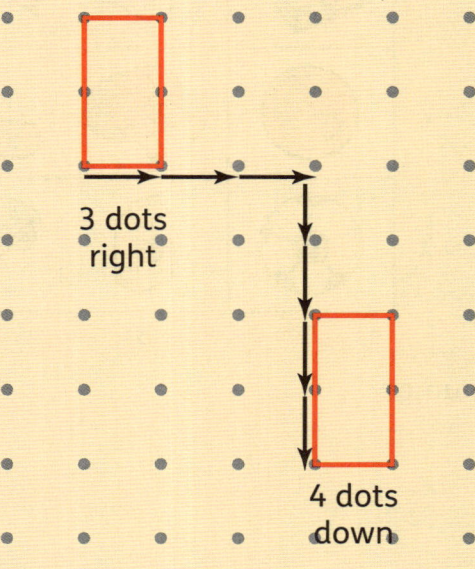

3 dots right

4 dots down

1 Read the directions for moving each shape. Copy these shapes on a dotty grid. Then draw their new positions.

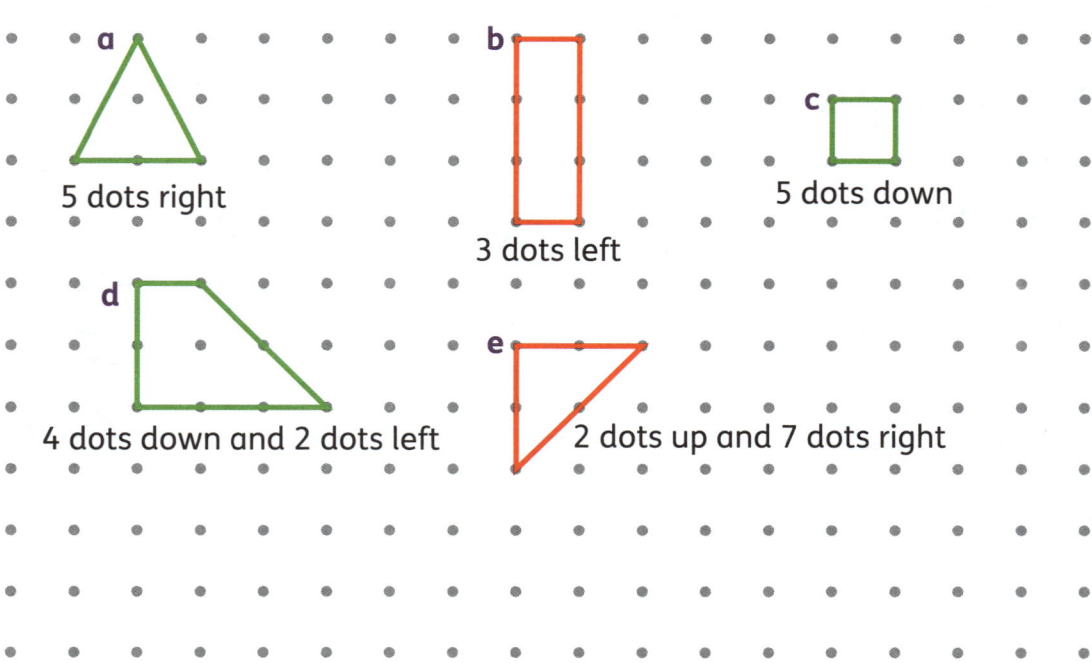

a 5 dots right

b 3 dots left

c 5 dots down

d 4 dots down and 2 dots left

e 2 dots up and 7 dots right

lesson continues ▶

2 Draw your own shape on a dotty grid.
Write instructions to move the shape on the grid.
Then draw the shape in its new position.

Problem solving

Use counters or small objects on the grid.

3 Start on A1. Follow these instructions:

- Move 4 blocks north.
- Move 2 blocks east.
- Move 2 blocks south.
- Move 2 blocks west.

What letter did you trace?

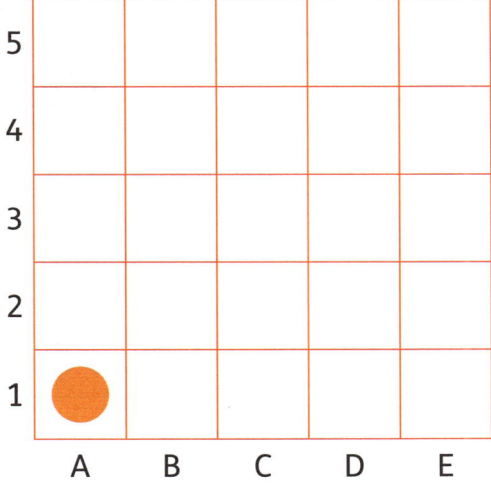

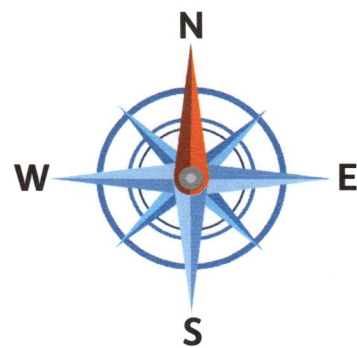

4 How could you trace these letters:

a L

b N

c V?

5 Write your own instructions for a partner. Let them follow your instructions. Did they finish in the correct position?

Fractions

Parts of a whole

Think and share

Which one doesn't belong? Why?

Talk about what you see.

How can you see it differently?

(There are no right or wrong answers!)

A B C D

 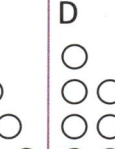

This shape is divided into four equal parts.

One-quarter of the shape is shaded purple. We write $\frac{1}{4}$.

Three-quarters of the shape is shaded green. We write $\frac{3}{4}$.

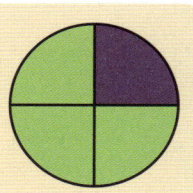

1 What **fraction** of each shape is shaded?

a    b 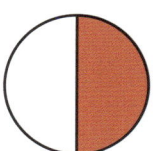   c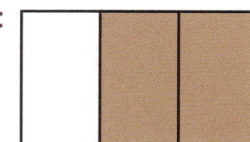

d    e    f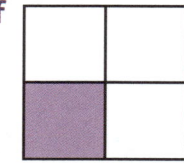

2 Draw your own shapes. Colour these fractions.

   a one-half, $\frac{1}{2}$ **b** two-thirds, $\frac{2}{3}$ **c** three-quarters, $\frac{3}{4}$ **d** four-fifths, $\frac{4}{5}$

Problem solving

3 Look at the shapes you have drawn.

   a What fraction of each shape is not coloured?

   b Tell a partner how you worked this out.

➡ *Workbook page 73*

More fractions

When we divide a whole object or group into equal parts, the equal parts are called **fractions**. We write a fraction with one number on top of another number. The top number is called the **numerator**. The bottom number is the **denominator**.

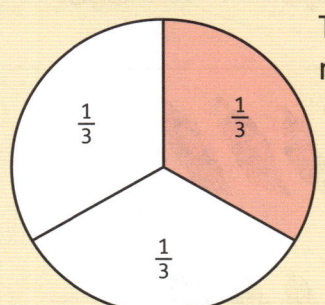

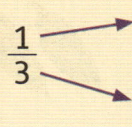

numerator – tells us how many parts in the fraction

denominator – tells us how many equal parts in the whole

The blue bar shows 1 whole. The green bars show one whole divided into three equal parts.

$1 \div 3 = \frac{1}{3}$. We say one-**third**.

2 green bars $= \frac{2}{3}$. We say two-thirds.

3 green bars = three-thirds or one whole. $\frac{3}{3} = 1$

Each green bar is $\frac{1}{3}$ or one out of three equal parts.

Can you see that $\frac{1}{3}$ is also a way to write $1 \div 3$?

1 Write the fraction of each shape that is shaded green.

a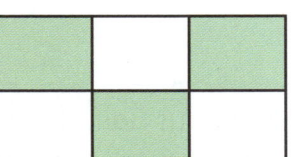

b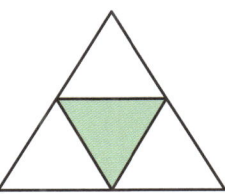

c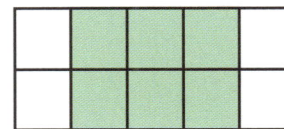

d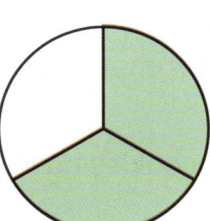

e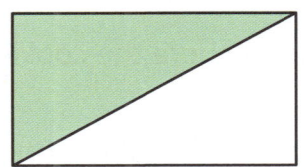

f

lesson continues ⊙

2 What fraction of each set is circled? Explain to a partner how you know.

a

b

c

d

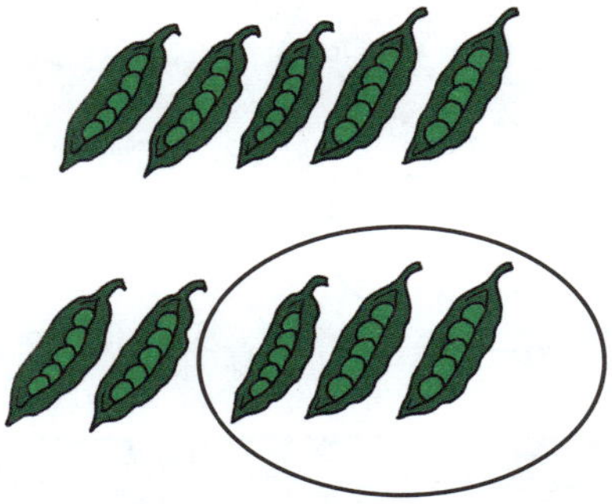

e

f

g

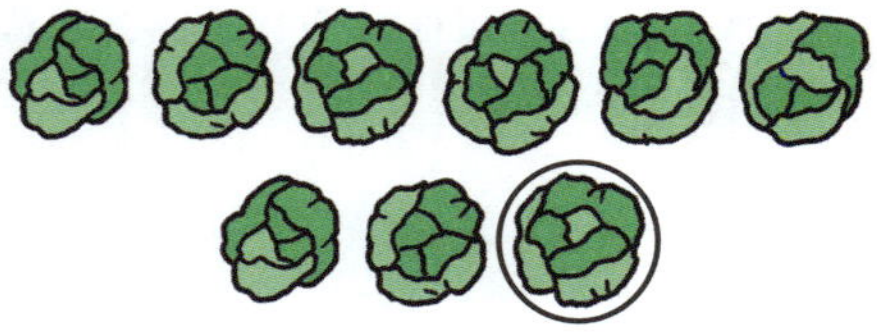

h

First work out half of the set. Then work out one-third of the remaining set.

Problem solving

3 I have a box of 24 cookies. I share the cookies with my brother so we each get half.

I eat $\frac{1}{3}$ of my share. What fraction of the original box of cookies have I eaten?

➡ *Workbook page 74*

Half of an amount

Here are six counters.

The counters have been divided into two equal groups.

Each group has half the counters. One **half** of 6 is 3.

We write: $\frac{1}{2}$ of 6 = 3

1 Use a line to divide each set of counters into **halves**.

Copy the number sentences and complete them.

a

$\frac{1}{2}$ of 4 = ☐

b

$\frac{1}{2}$ of 8 = ☐

c

$\frac{1}{2}$ of 10 = ☐

d

$\frac{1}{2}$ of 20 = ☐

2 How can you use the 2 times table to help you to work out half of a group?

Use counters or small objects to help you.

💡 Problem solving

3 16 children get on a bus. Half of the children get off at the first stop. At the next stop, half of the remaining children get off. How many children are left on the bus?

Half of an odd number

Jess baked 9 cookies.

She wants to give half of the cookies to a friend.

Jess asks: 'What is half of 9?'

Half of 8 is 4

So, half of 9 cookies is $4\frac{1}{2}$ cookies.

Half of 1 is $\frac{1}{2}$

To find half of an odd number, find half of the previous even number, then add $\frac{1}{2}$ (half of the extra 1).

1 What is half of each group of cookies?

a

b

c

d

2 Find:

   a $\frac{1}{2}$ of 29   b $\frac{1}{2}$ of 41   c $\frac{1}{2}$ of 25

   d $33 \div 2$ e $27 \div 2$ f $37 \div 2$

Find fractions

To find $\frac{1}{2}$ of an amount, we divide the amount into 2 equal groups.

- $\frac{1}{2}$ of 6 is the same as $6 \div 2 = 3$

To find any fraction of an amount, we divide the amount by the denominator of the fraction.

- $\frac{1}{3}$ of 6 is the same as $6 \div 3 = 2$

- $\frac{1}{4}$ of 8 is the same as $8 \div 4 = 2$

1 Copy and complete the number sentences.

a $\frac{1}{4}$ of 4 = ☐   b $\frac{1}{4}$ of 8 = ☐   c $\frac{1}{4}$ of 40 = ☐

d $\frac{1}{3}$ of 3 = ☐   e $\frac{1}{3}$ of 9 = ☐   f $\frac{1}{3}$ of 15 = ☐

2 Divide these cars into equal groups. Copy and complete the sentences.

a $\frac{1}{2}$ of ☐ = ☐   b $\frac{1}{4}$ of ☐ = ☐   c $\frac{1}{3}$ of ☐ = ☐

d $\frac{3}{4}$ of ☐ = ☐   e $\frac{2}{3}$ of ☐ = ☐

3 How much is each of these amounts of money?

a $\frac{1}{2}$ of 40c   b $\frac{1}{4}$ of 48c   c $\frac{1}{3}$ of 27c

4 Mary has a piece of rope that is 30 cm long.

a How long is half of the rope?   b How long is $\frac{1}{4}$ of the rope?

c Mary cuts the rope into two pieces. One piece is one-third of the length. How long is each piece of rope?

Problem solving

5 a What does $\frac{2}{3}$ mean?

b What is $\frac{2}{3}$ of 15?

c How can you find $\frac{3}{4}$ of 40?

Use number rods or number frames to help you, or draw a sketch.

▶ *Workbook page 75 and page 76*

Equivalent fractions

Equivalent fractions have the same value.

 $= \frac{1}{2}$  $= \frac{2}{4}$ 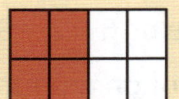 $= \frac{4}{8}$

1 Write the equivalent fractions.

a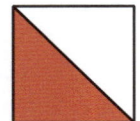

$\frac{1}{2}$ = $\frac{\square}{8}$

b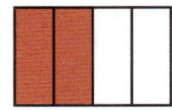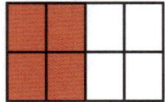

$\frac{2}{4}$ = $\frac{\square}{8}$

c

$\frac{5}{10}$ = $\frac{\square}{2}$

d

$\frac{1}{2}$ = $\frac{\square}{4}$

2 Which diagrams show equivalent fractions?

A 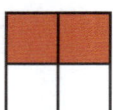 **B**  **C**  **D**

E 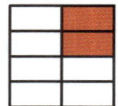 **F** 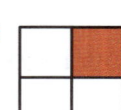 **G** 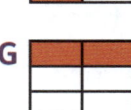 **H**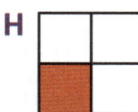

Problem solving

3 Which shapes make equivalent fractions in this picture? Discuss with a partner and draw sketches to show your answers.

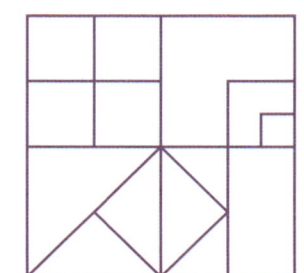

Tenths

When we divide an object or set into 10 equal parts, the fractions are called **tenths**.

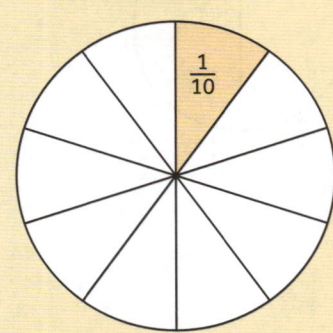

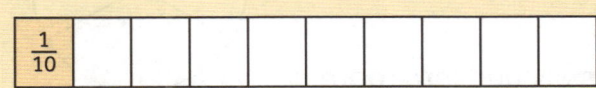

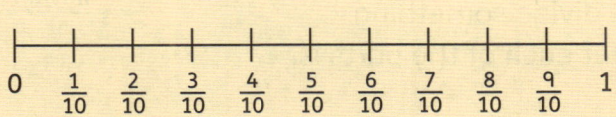

1 How many tenths are shaded in each picture? Write your answer as a fraction.

a

b

c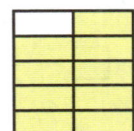

d

e

f

2 Use the pictures to help you work out these fractions of amounts.

a $\frac{3}{10}$ of 20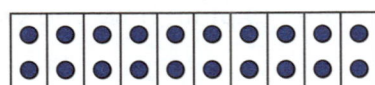

b $\frac{5}{10}$ of 30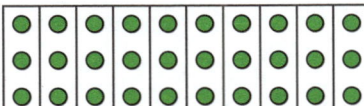

c $\frac{1}{10}$ of 40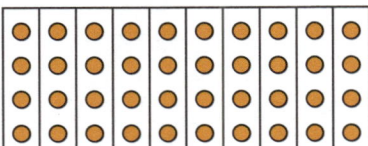

3 Use any of the diagrams on this page to help you work out these equivalent fractions.

a $\frac{1}{2} = \frac{\square}{10}$

b $\frac{3}{5} = \frac{\square}{10}$

c $1 = \frac{\square}{10}$

Compare fractions

We can compare fractions using <, > or =.

$\frac{1}{3}$ ☐ $\frac{1}{2}$

The first circle is divided into more parts than the second circle. The more parts we divide something into, the smaller each of the parts is.

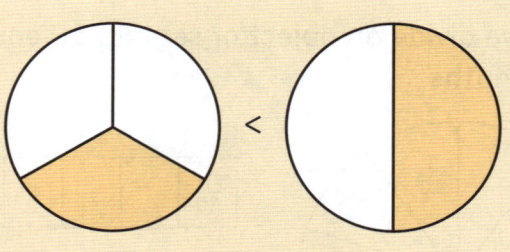

$\frac{1}{3}$ is less than $\frac{1}{2}$.

1 Compare the fractions using <, > or =. Use the diagrams to help you.

a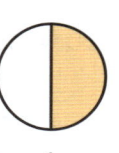

$\frac{1}{4}$ ☐ $\frac{1}{2}$

b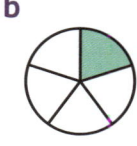

$\frac{1}{5}$ ☐ $\frac{1}{6}$

c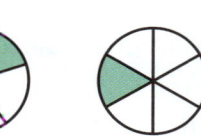

$\frac{1}{2}$ ☐ $\frac{2}{4}$

d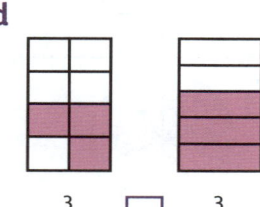

$\frac{3}{8}$ ☐ $\frac{3}{5}$

e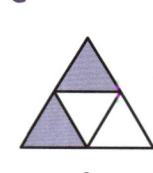

$\frac{2}{4}$ ☐ $\frac{1}{4}$

f

$\frac{6}{10}$ ☐ $\frac{3}{5}$

2 Compare the fractions using <, > or =. Explain how you worked each answer out.

a $\frac{3}{4}$ ☐ $\frac{1}{8}$ b $\frac{4}{5}$ ☐ $\frac{4}{10}$ c $\frac{2}{3}$ ☐ $\frac{4}{6}$

3 $\frac{1}{5}$ is greater than $\frac{1}{2}$ because 5 is greater than 2.

What is this person's mistake?

4 Why is $\frac{1}{2}$ of 6 less than $\frac{1}{2}$ of 10?

5 Compare amounts using <, > or =.

a $\frac{1}{3}$ of 9 ☐ $\frac{1}{2}$ of 4   b $\frac{1}{5}$ of 20 ☐ $\frac{1}{2}$ of 8

c $\frac{1}{4}$ of 12 ☐ $\frac{1}{2}$ of 6

You can use groups of objects to help you work these out.

Fractions of a number

1 This picture shows a large cube made of small cubes.

It is made of rows of purple cubes and rows of green cubes.

Use the picture or make your own cube to help you answer the questions.

   a How many small cubes are there altogether?

   b If you take one-third of the cubes away, how many cubes are left?

   c Draw $\frac{1}{3}$ of the cubes.

   d $\frac{2}{3}$ of the cubes are in one pile. The rest are in another pile.

     How many cubes are there in each pile?

2 This picture shows another large cube made of small cubes.

It is made of rows of green cubes and rows of purple cubes.

   a How many small cubes are there altogether?

   b What fraction of the cubes are green?

   c Draw $\frac{1}{8}$ of the purple cubes.

   d Add together $\frac{1}{4}$ of the purple cubes and $\frac{1}{4}$ of the green cubes. How many cubes do you have?

   e Put $\frac{1}{4}$ of the green cubes in a row. Put $\frac{3}{8}$ of the purple cubes in a row.

     Which row has more cubes?

3 **a** Make or draw your own cube or cuboid. Try a different size or use three colours for the smaller cubes.

   b Think of some fraction questions about your cube or cuboid to ask a partner.

Add and subtract fractions

A pizza is divided in quarters. Vikal has three of the quarters. He gives one of his quarters to his brother. How much does Vikal have left?

$$\frac{3}{4} - \frac{1}{4} = \frac{2}{4}$$

Vikal has $\frac{2}{4}$ left. $\frac{2}{4}$ is equivalent to $\frac{1}{2}$, so we can also say he has half a pizza left.

When denominators are the same, you can add or subtract numerators.

Kayli has $\frac{2}{3}$ of a box of juice. Liana has $\frac{1}{3}$ of a box of juice. How much juice do they have altogether?

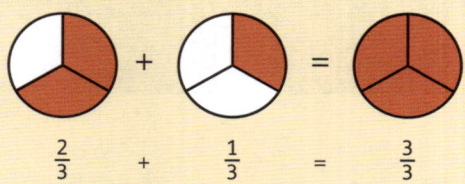

$$\frac{2}{3} \quad + \quad \frac{1}{3} \quad = \quad \frac{3}{3}$$

They have three-thirds or 1 whole box of juice.

1　Use the diagrams to help you add or subtract.

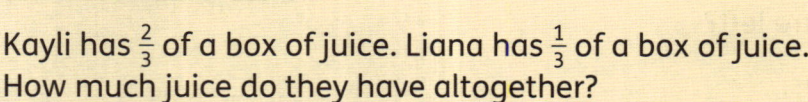

a
$$\frac{1}{2} \quad + \quad \frac{1}{2}$$

b
$$\frac{1}{3} \quad + \quad \frac{1}{3}$$

c
$$\frac{3}{4} \quad - \quad \frac{1}{4}$$

d
$$\frac{7}{8} \quad - \quad \frac{6}{8}$$

e
$$\frac{2}{5} \quad + \quad \frac{1}{5}$$

f
$$\frac{1}{8} \quad + \quad \frac{2}{8}$$

 Problem solving

2　What fraction of the circle is shaded and what fraction is not shaded?

How did you work it out?

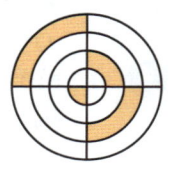

➡ *Workbook page 77*

UNIT 16

Capacity and temperature

Measure capacity

💭 Think and share

With water from a hose, can you measure exactly
4 **litres** using these two buckets?

What other litre amounts can you make?

> You are allowed to fill and
> empty the buckets as
> many times as you want.

3 litres

5 litres

We can use litres to measure how much a large container holds.

We use millilitres to measure how much a small container holds.

There are 1000 millilitres in 1 litre.
$1\,\ell = 1000\,ml$

1. Measure and record how much liquid you drink in a day.

   a Make a class chart to show how many millilitres of liquid each pupil in the class drinks in a day.

   b Do the pupils drink more than 1 litre or less than 1 litre each day?

2. Make a list of 10 containers that you use at home to hold liquids.

   a Write the containers in order, from the container that holds the least to the container that holds the most.

   b What do you use each container for? Tell a partner.

➡️ *Workbook page 78*

Litres and millilitres

1 teaspoon holds 5 millilitres.

One litre equals one thousand millilitres.

1 ℓ = 1000 ml

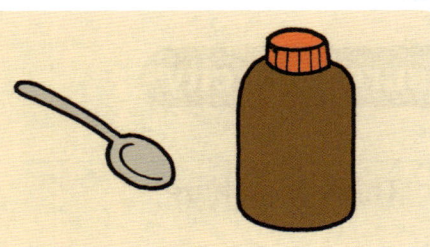

1 Would you measure these in millilitres or litres?

a  petrol in a can	**b**  one serving of yoghurt	**c**  a bottle of medicine
d 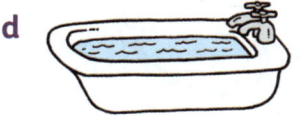 water in a bath	**e** 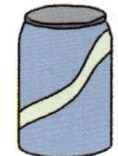 drink in a can	**f**  drink in a large bottle

2 Choose the best estimate for each container.

a

100 ml or 10 ℓ

b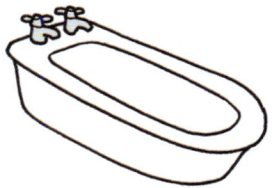

100 ml or 100 ℓ

c

250 ml or 250 ℓ

 Problem solving

3 How could you use a 1-litre container and some sand to find out whether another container holds more or less than a litre?

Read scales

In this set of jugs there is a total of 2 litres and 200 ml of liquid.

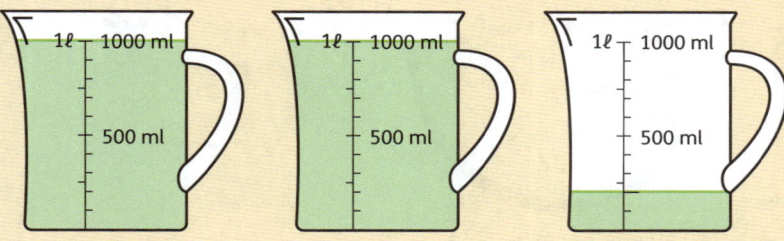

We can write this as 2 ℓ 200 ml or 2200 ml.

1 How much liquid is there in each set of jugs? Write each answer in two ways.

a

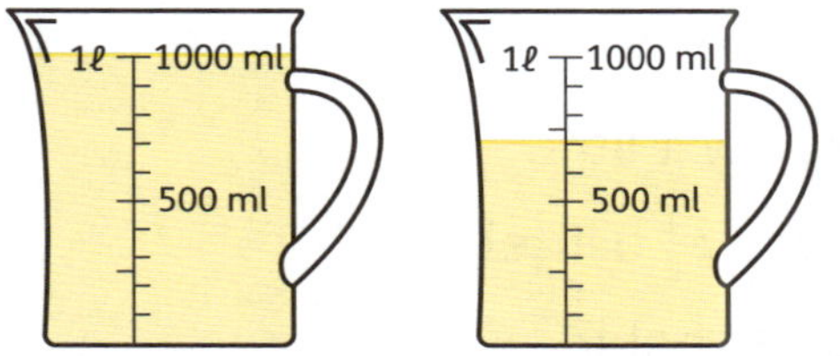

b

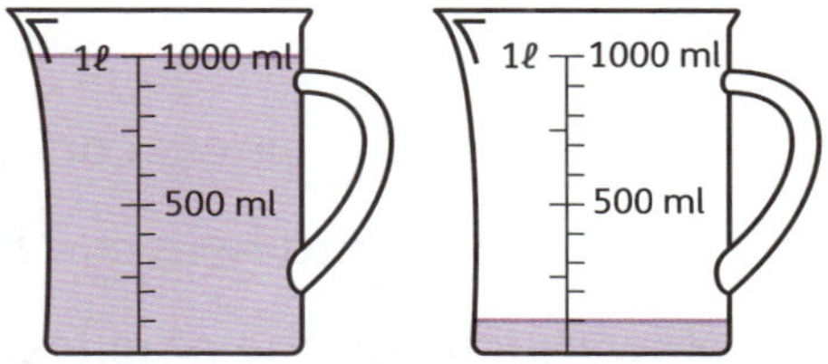

c

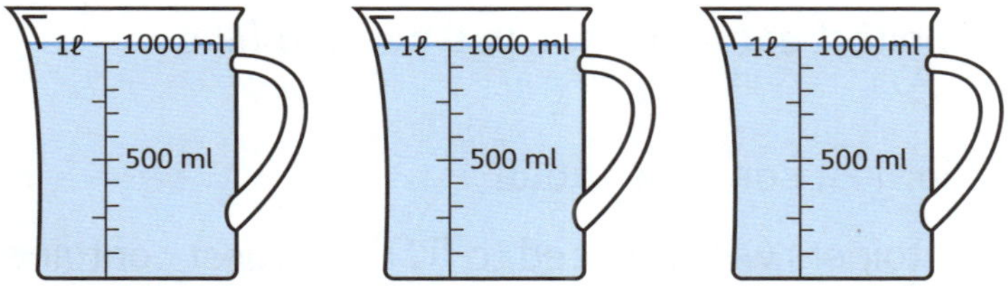

d

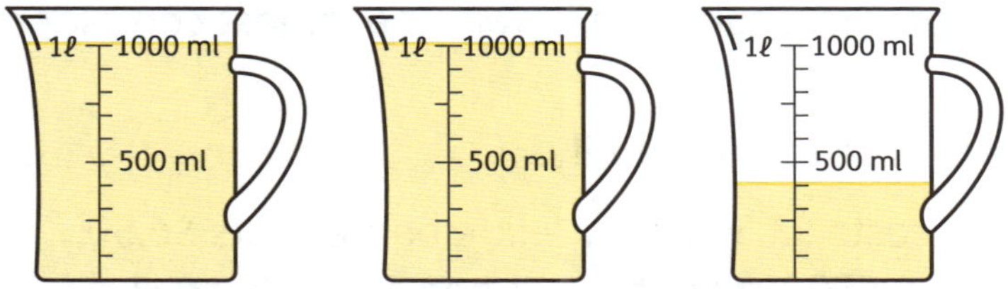

e

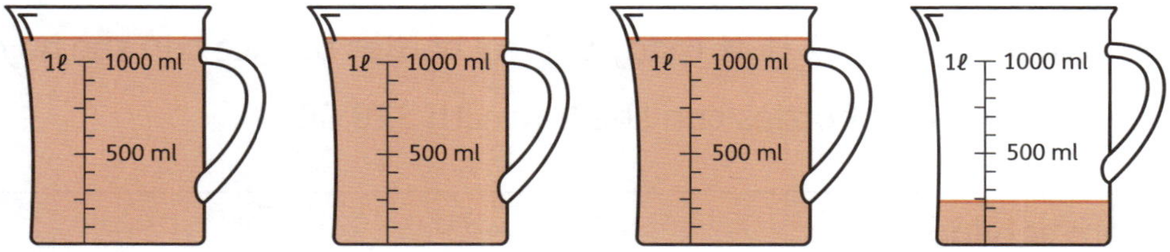

2 Now round each amount from question 1 to the nearest litre.

3 How much will be left from a full 1-litre jug if you pour out:

a 100 ml   b 400 ml   c half of the liquid?

Capacity problems

1 Look at the containers and answer the questions.

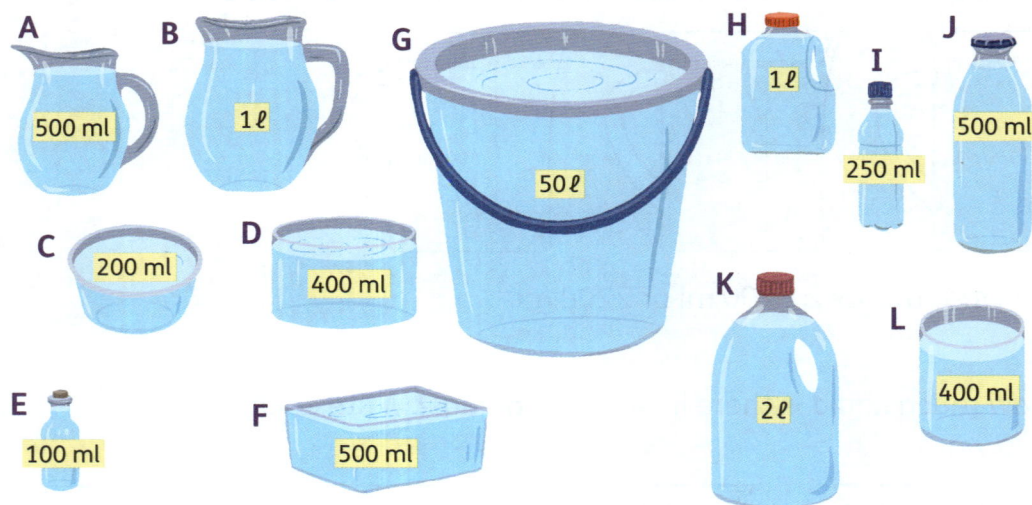

a How many containers can hold exactly 1 litre?

b How many containers can hold exactly $\frac{1}{2}$ a litre?

c How much can the smallest container hold?

d How many of the smallest container will you need to fill a
   1-litre container?

e How much can the largest container hold?

f How many 2-litre containers will you need to fill the largest container?

g How many $\frac{1}{2}$-litre containers can you fill from the largest container?

2 Mrs Singh wants to buy juice for her family.

A 500-ml bottle of juice costs $3.50 and a 1-litre bottle costs $6.50.

a Is it cheaper to buy two 500-ml bottles or one 1-litre bottle?

b Mrs Singh has $20.00. How many 500-ml bottles of juice can she buy?

c How many 1-litre bottles can she buy with $20.00?

Problem solving

3 Tomas has a 2-litre bottle of juice. He drinks 200 ml of juice each day for
4 days. How much juice does he have left at the start of day five?

➡️ *Workbook page 79*

Temperature

When we measure **temperature**, we measure how hot something is. We use a **thermometer** to measure temperature. The units are called **degrees**.

This thermometer shows a temperature of 40 **degrees Celsius**. We write it like this: 40 °C.

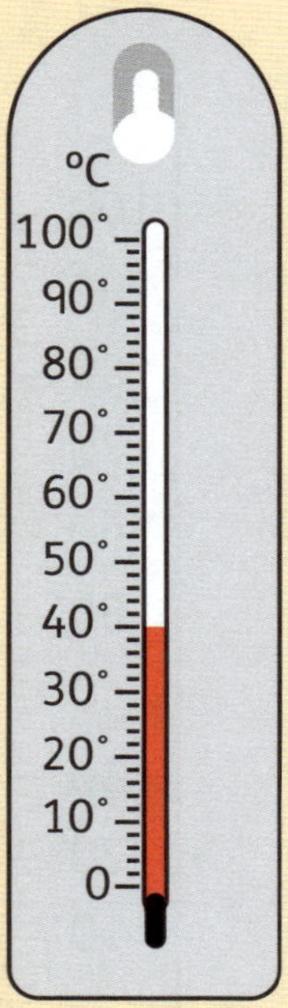

°C
100°
90°
80°
70°
60°
50°
40°
30°
20°
10°
0

Water freezes at 0 °C.

Between these temperatures, water is liquid.

Water boils and turns to steam at 100 °C.

1 Write these temperatures using the degree symbol. The first one is done for you.

a 2 degrees more than 30 degrees Celsius   32°C

b 30 degrees more than 30 degrees Celsius

c 10 degrees more than the freezing point of water

d 20 degrees less than the boiling point of water

e Halfway between the freezing point and the boiling point of water

lesson continues ▶

2 Write the temperature shown on each thermometer.

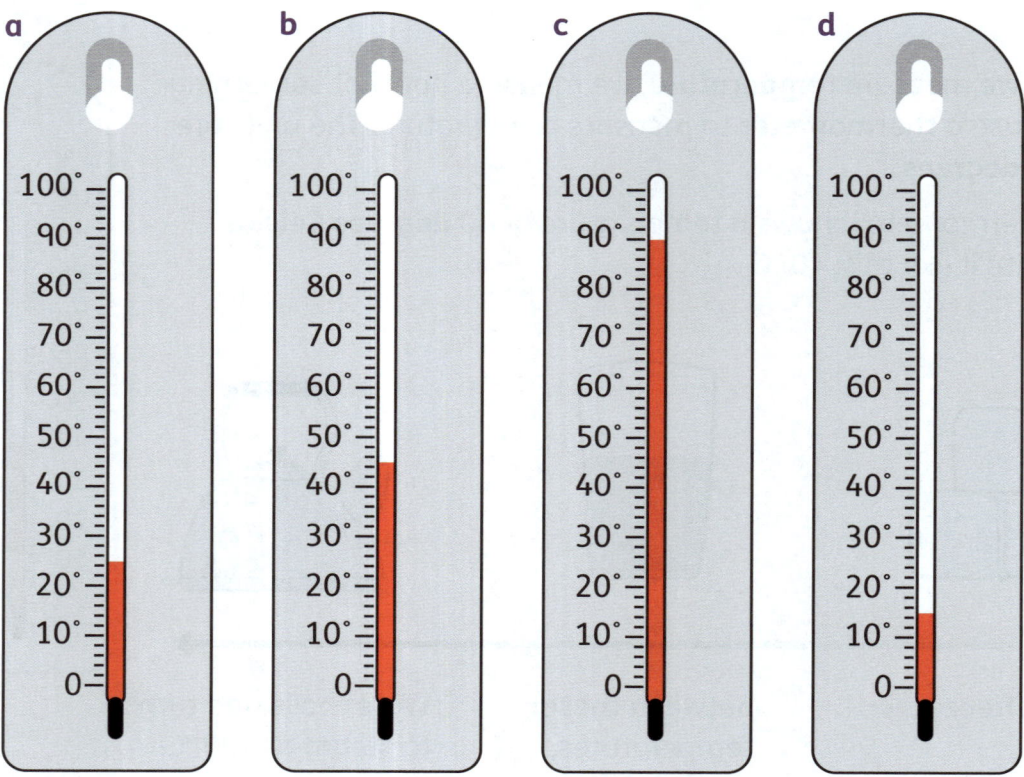

3 Some pupils measured the air temperature in their school playground for one week. They recorded the highest and lowest temperatures in this table.

	Mon	Tue	Wed	Thur	Fri	Sat	Sun
Highest temperature	28 °C	29 °C	32 °C	26 °C	23 °C	32 °C	35 °C
Lowest temperature	25 °C	26 °C	29 °C	25 °C	22 °C	29 °C	28 °C

a Which was the hottest day?

b Which was the coolest day?

c Which two days had the same highest and lowest temperatures?

d On which day was the difference between the hottest and the coolest temperatures greater than 3 degrees?

e Could these temperatures be from your country? How do you know?

4 Find the temperatures for this week in a newspaper weather forecast, or a weather app. Draw a table showing the expected temperatures.

➡ *Workbook page 80*

Temperature experiments

You will need:

cups

jugs

some ice

a kettle

a thermometer

a tap

an adult to help you

a fridge

CAUTION: Boiling water can burn. An adult must help you pour the water from the kettle.

1 **a** Pour some water from the cold tap into a cup. Measure and record the water temperature.

b Next, do the same, but with water from the warm tap.

2 Estimate the temperature of the water if you:

a add ice to the cold water and wait one **minute**

b add ice to the warm water and wait one minute.

Now measure to test your estimates.

3 Your teacher will pour some just-boiled water into a cup. Measure the temperature of the water:

a immediately   **b** after 1 minute

c after 2 minutes   **d** after 5 minutes.

Make a table in your notebook to record all the temperatures.

4 Do you agree with this person? Why or why not?

> The boiled water keeps getting cooler the longer I leave it. If I leave the water for long enough, it will freeze.

➡ *Workbook page 81*

Probability

Will it happen?

☁ Think and share

When we talk about events that might happen, we use words like possible, likely and impossible.

It is impossible that the sun will be blue today.

It is possible that it will rain today.

With your class, discuss things that might or might not happen today.

- What will definitely happen?
- What is impossible? (It can't happen.)
- What might happen, but you aren't certain?

1 We use many different words to describe possible events.

a Match the words in boxes to the letters on the line.

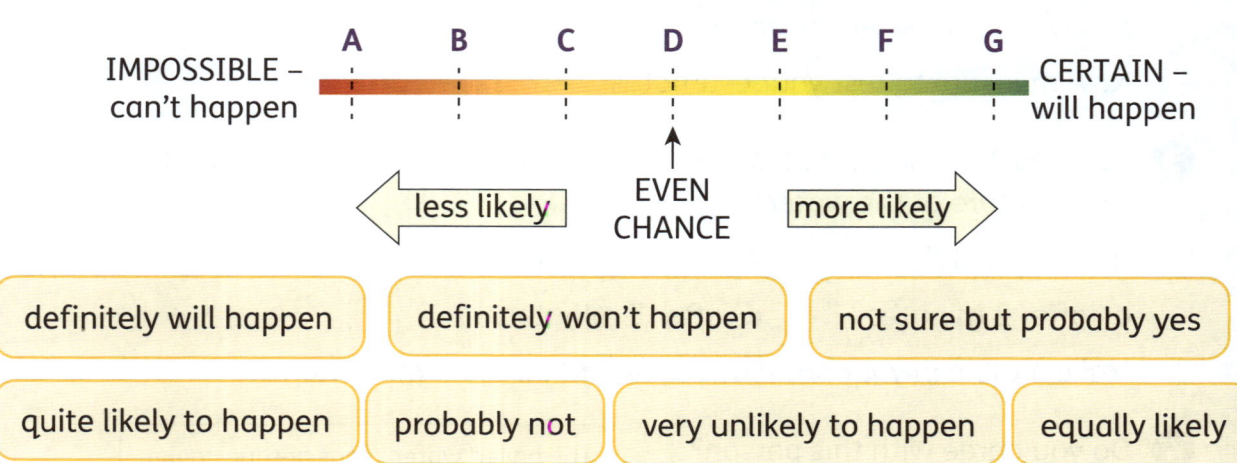

IMPOSSIBLE – can't happen

A B C D E F G

CERTAIN – will happen

less likely | EVEN CHANCE | more likely

| definitely will happen | definitely won't happen | not sure but probably yes |

| quite likely to happen | probably not | very unlikely to happen | equally likely |

b Tell your partner something that matches each box.

Possible outcomes

When you spin the arrow on this spinner, it could land on yellow, purple or green. There are three possible **outcomes**:

Outcome 1: Lands on yellow

Outcome 2: Lands on green

Outcome 3: Lands on purple

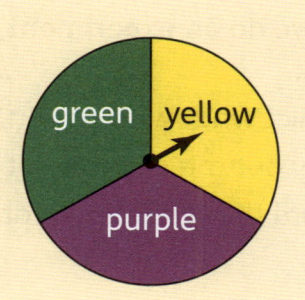

1 Look at each spinner. How many possible outcomes are there? What are the outcomes?

a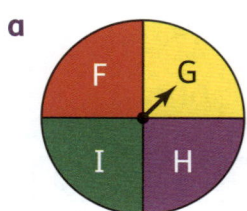

b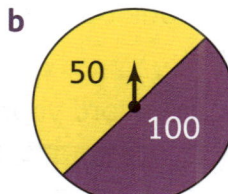

c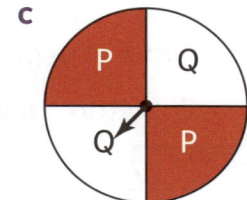

d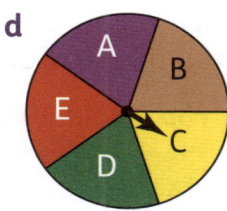

2 Look at these spinners. One outcome is more likely than the others. What is the most likely outcome for each spinner?

a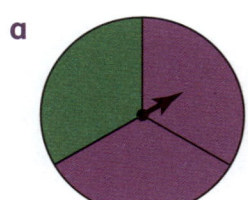

b

c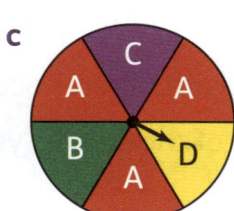

d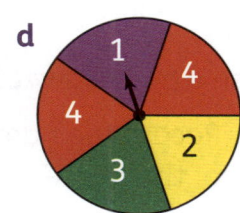

3 There are six possible outcomes when we spin a 1–6 spinner. What are they?

4 Think about what might happen when you spin a 1–6 spinner. Describe the **probability** of each statement below.

> You can use these words: *possible, certain, impossible, likely, very likely, small chance, great chance, even chance.*

a You will get *one* of these numbers: 1, 2, 3, 4, 5 or 6.

b You will get *none* of these numbers: 1, 2, 3, 4, 5 or 6.

c You will get an even number.

d You will get a number that starts with t or f.

e You will get a number greater than 1.

➡ *Workbook page 82*

Probability experiments

When we do an **experiment**, we try something out to see what happens.
We can:

- predict what will happen
- carry out the experiment
- record the results in a table.

1 When we flip a coin:

- there are two possible outcomes — heads or tails
- each outcome has an equal chance of happening.

But, if we flip a coin 10 times, we may get one outcome more often than the other. Try this experiment.

a Flip a coin 10 times. Copy this table and record your outcomes.

Flip	1	2	3	4	5	6	7	8	9	10
Outcome										

b Which outcome did you get the most often?

c Compare your results with some of your classmates.

2 Use a bag of sweets with different colours.

a Record the name of the sweets you chose. Draw a picture of the packet.

b Find the mass of the sweets. You can weigh them or look on the packet.

c Open the packet and count how many sweets there are altogether. Save the packet.

d Count the number of sweets of each colour. Record the numbers in a table.

e Look at the numbers for each colour. Record which colour occurs the most and which colour occurs the least.

f You are going to take 10 sweets, without looking. Predict how many sweets of each colour you might take.

g Now do the experiment. Put all the sweets back in the packet. Without looking, take 10 sweets. Record the actual outcome.

➡ *Workbook page 83*

Time

Reading time

💭 **Think and share**

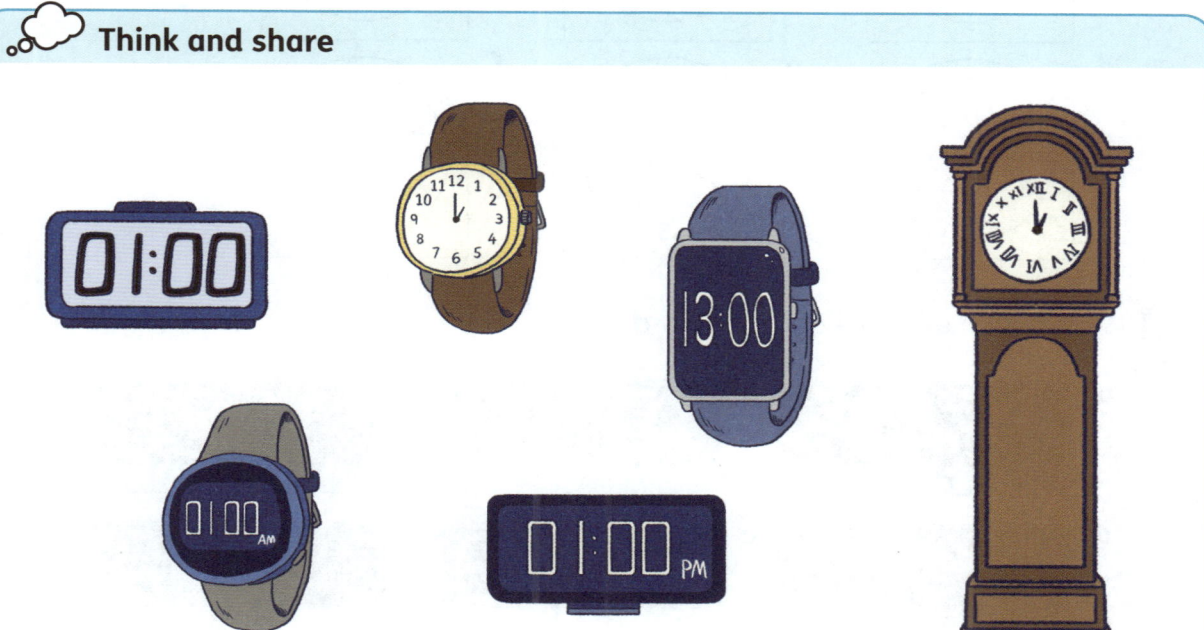

Which clocks show exactly the same time of day?

Which clocks could show different times of day? How do you know?

What will the time be one **hour** after the time shown? How do you know?

What else do you notice? What do you wonder?

---

1   **a**   Which clock does not show the same time as the others?

    b   What time is shown on that clock?

    c   Draw your own clocks to show 7 o'clock in three different ways.

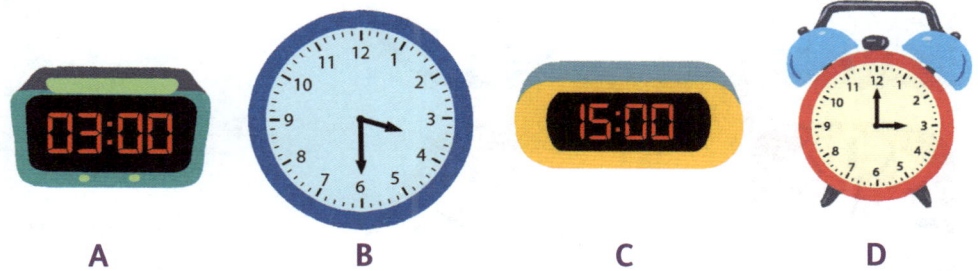

A   B   C   D

lesson continues 🔶

2 Write the time shown on each clock.

a 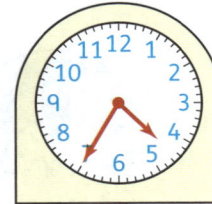   b   c

d   e   f

3 Write each of these times in words.

a    b    c

d    e    f

4 Draw **digital clock** faces to show each of these times.

a twenty to five   b five to twelve

c quarter to eight   d fifty-five **minutes** past three

5 Write these times using digital notation. So instead of half past one, you write 1:30.

a    b    c

d    e    f

➡ *Workbook page 84*

Analogue clocks

A clock that has a face and hands that point to the hours and the minutes is called an **analogue clock**. Here are some different analogue clocks:

Some clocks have lines or dots instead of numbers. You have to work out the time from the positions of the hands.

Some clocks use **Roman numerals**. This is an ancient number system. Here are the Roman numerals for the numbers 1 to 12:

1	2	3	4	5	6	7	8	9	10	11	12
I	II	III	IV	V	VI	VII	VIII	IX	X	XI	XII

1 Discuss the Roman number system with a partner.

   a Which numbers have their own letter?

   b Which numbers are made using the letters from other numbers?

   c How do you think the Roman number system works?

2 Write the Roman numerals for these numbers.

   a 13   **b** 15   **c** 20

3 In Roman numerals, L means 50. Work out the values of these Roman numerals.

   a XIII   **b** XXIV   **c** LXV

4 In Roman numerals, C means 100. What do these Roman numerals represent?

   a CC   **b** CCCIII   **c** CCI

lesson continues ⊙

5 What time is shown on each clock?

a

b

c

d

e

f

6 Look at this compass showing the cardinal points north, south, west and east. Which hour on an analogue clock is located at each point?

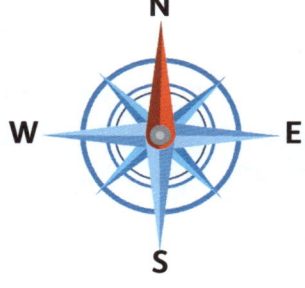

a north   **b** west

c south   **d** east

 Try it out with some different times on an analogue clock.

💡 **Problem solving**

7 Vusi says to his friend: 'I can work out the minutes past the hour. I look at the number the minute hand is pointing to and multiply this number by 5.'

- Do you agree with Vusi?

- Why does this work?

- Are there some times that it doesn't work?

Digital times

Digital clocks display numbers to show the time.

a 5:05   b 10:22   c 4:26
d 2:25   e 9:59   f 11:50

The numbers before the colon (:) show us the hour.

The next two numbers show how many minutes have passed after the hour. The colon separates the hour and the minutes.

There are different kinds of digital clocks.

- The **12-hour clock** only uses the numbers 1 to 12 for the hours. It may also show a.m. for morning times and p.m. for evening times.

- The **24-hour clock** uses the numbers 0 to 23 for the hours.

  The day starts at 00:00, midnight. 00:01 is one minute after midnight. 12:00 is noon or midday. 13:00 is 1:00 p.m. 23:00 means 11:00 p.m.

  After 23:59, the time changes to 00:00 again, which is midnight.

1 These digital clocks all show times in the afternoon or evening. Write the correct time in words. The first one is done for you.

a 14:45   b 13:25   c 16:10   d 19:30

quarter to three

e 22:35   f 20:50   g 23:05   h 18:55

2 Is 00:05 a morning time or an evening time? Give reasons for your answer.

3 Write these times in 24-hour clock notation.

a 1:15 a.m.   b 7:15 p.m.   c half past eleven at night

d half past six in the evening   e 25 to ten in the morning

lesson continues ➤

4 Answer these questions about the flight departures board.

   a Which is the earliest flight?

   b Which flight is at ten to four in the afternoon?

   c How many flights are there between 5 pm and 7:30 pm?

   d There is a flight at quarter to one. Where does it fly to?

   e Judging by the gates that are open, about what time do you think
   it was when the board showed this information?

   f How long before the departure time does the gate open? How did
   you work it out?

   g TBA means 'To Be Announced'. Work out the gate opening times
   for the flights to Accra and Colombo.

✈ DEPARTURES

Time	Destination	Gate status	Gate opens
11:30	DUBAI	Closed	--
12:45	BANGKOK	Closed	--
14:10	NAIROBI	Closed	--
15:20	KUALA LUMPUR	Open – A9	14:40
15:50	NEW DELHI	Opening soon – A7	15:10
16:30	CAIRO	A3	15:50
17:05	MANAMA	A5	16:25
18:00	ACCRA	TBA	TBA
18:05	COLOMBO	TBA	TBA
19:20	ISLAMABAD	TBA	TBA
20:10	MOSCOW	TBA	TBA

5 A flight departs at 23:30 in the evening. The flight takes $7\frac{1}{2}$ hours.
What time does it land?

6 Make up your own three questions about flight departure times.
You can use the board above, or make up your own times.
Ask a partner to solve your questions.

➡ *Workbook page 85*

Units of time

Some things take much longer than others.

We use many different units of time to measure how long things take.

It took me one **second** to make this dot.

My painting took an hour to finish.

This seedling grew in 3 **days**.

I planted this tomato plant 2 **months** ago.

1 minute = 60 seconds

1 **week** = 7 days

1 hour = 60 minutes

1 month = about 4 weeks

1 full day = 24 hours

1 **year** = 12 months

1 Write five things that you do every day.

a Estimate how long it takes you to do each activity.

b Time yourself as you do each thing today.

Record how long you took to the nearest minute.

c How close were your estimates to your actual times?

lesson continues ⟩

2 Look at these activities. How long do you think each activity takes?
Choose and write the best estimate.

a Eat a meal

30 seconds
5 minutes
30 minutes
5 hours

b Read two pages
of a story book

10 seconds
10 minutes
50 minutes
10 hours

c Drink a glass
of water

15 seconds
10 minutes
15 minutes
15 hours

d Have a bath

5 seconds
50 seconds
20 minutes
3 hours

e Open a parcel

1 second
30 seconds
30 minutes
30 hours

f Change into your
sports clothes

5 seconds
15 seconds
5 minutes
5 hours

➡ Workbook page 86

Time and time intervals

On Monday, when Jenna starts watching TV, the time is 6 o'clock. When she finishes watching TV, the time is half past 7.

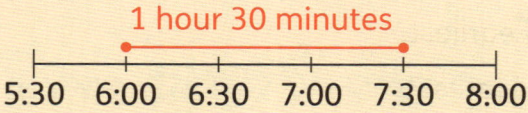

1 hour 30 minutes

5:30   6:00   6:30   7:00   7:30   8:00

Jenna watches TV for a **time interval** of $1\frac{1}{2}$ hours. A time interval is the amount of time that passes from one time to another. It is also called a period of time.

1. Jenna watches TV for one and a half hours each day. These are the times she starts watching. At what time does she finish watching TV each day?

   a   Tuesday: 5 o'clock

   b   Wednesday: 4:15 p.m.

   c   Thursday: 6:45 p.m.

2. Jenna's little brother is allowed 30 minutes of TV each day. These are the times he finishes watching TV. At what time did he start watching TV each day?

   a   Monday: 6:30 p.m

   b   Tuesday: 4:15 p.m.

   c   Wednesday: 5:00 p.m.

3. Work out the length of each time interval.

   a

   7:00   7:30   8:00   8:30   9:00   9:30

   b
   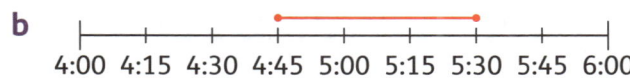

   4:00 4:15 4:30 4:45 5:00 5:15 5:30 5:45 6:00

lesson continues ◗

4 Four teams took part in a treasure hunt. They started at different times. They finished at different times. The table shows their start and finish times.

Team	Start time	Finish time
Red	10:20	11:35
Blue	10:25	11:30
Green	10:30	11:40
Yellow	10:35	11:45

a Which team started earliest?

b Which team finished latest?

c Which team started at half past ten?

d Which team finished at quarter to twelve?

e Which team took the longest time?

f Which team took the shortest time?

g In the next treasure hunt, the teams start 10 minutes later. What time does each team start?

h The teams finish five minutes earlier. What time does each team finish?

5 Kayley wrote a list of tasks for the weekend.

help hang up laundry – 20 minutes

go for a run – 30 minutes

finish my science project – 1 hour

listen to my favourite podcast – 40 minutes

practise keyboard – 35 minutes

Copy and complete this table to show her starting and finishing times for each activity.

Task	Start time	Finish time
Laundry	11 a.m.	
Run	9:15 a.m.	9:45 a.m.
Science project	4:35 p.m.	
Podcast	6:20 p.m.	
Keyboard practice		2:15 p.m.

➡ *Workbook page 87*

Calendars and dates

Use this calendar page to answer the questions.

August						
Mon	Tue	Wed	Thu	Fri	Sat	Sun
			1	2	3	4
5	6	7	8	9	10	11
12	13	14	15	16	17	18
19	20	21	22	23	24	25
26	27	28	29	30	31	

1 Write the date of:

   a the first Sunday in August

   b the last Sunday in August

   c the third Friday in August.

2 Write all the dates for one week, starting with:

   a the 3rd of August   **b** the 13th of August

3 What day of the week is each date?

   a the 1st of August   **b** the 17th of August   **c** the 30th of August

4 Write the day and the date five days after each date.

   a the 10th of August   **b** the 23rd of August   **c** the 26th of August

5 What is the day and date 14 days before each date?

   a the 27th of August   **b** the 17th of August

---

6 For each date, write the day of the week.

   a 10th January   **b** 1st March

   c 8th May   **d** 4th August

   e 9th October   **f** 31st December

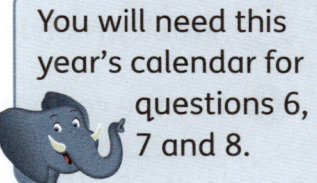

You will need this year's calendar for questions 6, 7 and 8.

7 Write the date of each day.

   a the first Sunday in April   **b** the second Friday in September

   c the third Wednesday in July   **d** the last Thursday of November

8 What date is your birthday? What day of the week is your birthday this year?

➡ *Workbook page 88*

Use these calendar pages for February, March and April to answer the questions.

February

M	Tu	W	Th	F	Sa	Su
					1	2
3	4	5	6	7	8	9
10	11	12	13	14	15	16
17	18	19	20	21	22	23
24	25	26	27	28		

March

M	Tu	W	Th	F	Sa	Su
					1	2
3	4	5	6	7	8	9
10	11	12	13	14	15	16
17	18	19	20	21	22	23
24	25	26	27	28	29	30
31						

April

M	Tu	W	Th	F	Sa	Su
	1	2	3	4	5	6
7	8	9	10	11	12	13
14	15	16	17	18	19	20
21	22	23	24	25	26	27
28	29	30				

1 Sandra visits her grandparents every two weeks. Her last visit was on 8th February. When will she visit them again?

2 Ishmael gets his pocket money on the last Friday of each month. On which dates did he get his pocket money during these three months?

3 **a** On 1st March, Indira's friend invites Indira to a party on 15th March. How many school days are there from 1st March to 15th March?

b One week before the party, Indira buys a new dress for the party. What date is this?

Problem solving

4 Use the calendar for April. Read the clues to work out each date.

a	**b**
• This date is an odd number.	• This date is an odd number.
• It is in the fourth week of the month.	• The tens digit is an even number.
• It is in the 5 times table.	• It is in the last three days of the month.

Mixed practice 3

1 Which of these objects is not shaped like a cuboid?

> a cereal box   a book   a box of matches   a straw

2 Is each statement true or false? If the statement is false, write the correct statement.

   a A vertex is the same as a corner.

   b A cube has 4 faces.

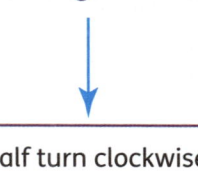

3 The compass shows you the points north, west, south and east. Which direction will each arrow face after the turn?

   a

   b

   c

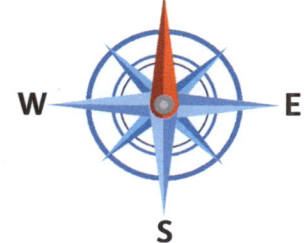

> a quarter turn clockwise   a quarter turn anti-clockwise   a half turn clockwise

4 Which shape does not match the others?

   a 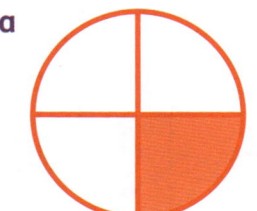   **b**   **c**   **d**

5 What fraction of each diagram is shaded?

   a 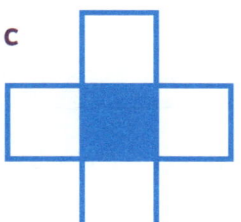   **b**   **c**   **d**

6 Each of these fractions is incorrect. Explain why.

   a 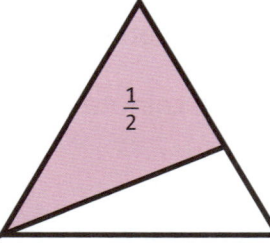 $\frac{1}{2}$

   b  $\frac{1}{4}$

   c $\frac{1}{5}$

7 Bella saves \$32. She spends $\frac{1}{4}$ of her money on a watch.

   a How much does she spend?

   b What fraction of the money does she have left?

8 Work out:

   a $\frac{1}{5}$ of 25   **b** $\frac{1}{4}$ of 20   **c** $\frac{1}{3}$ of 18

9 Copy and complete this number line. Label all the marks with the correct fractions.

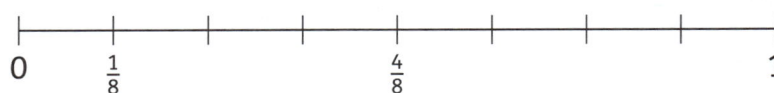

$\begin{array}{llll} 0 & \frac{1}{8} & \frac{4}{8} & 1 \end{array}$

10 The long stick is about 1 metre long. Estimate the length of the ribbon.

1 m

11 Work out:

   a $\frac{2}{5} + \frac{1}{5} = \square$ **b** $\frac{4}{10} + \frac{3}{10} = \square$ **c** $\frac{9}{10} - \frac{4}{10} = \square$

12 Salvina has made a mistake in this addition.

$\frac{1}{8} + \frac{3}{8} = \frac{4}{16}$

Explain her mistake.

13 Write these capacities in order from smallest to greatest.

$\frac{3}{4}$ litre   500 ml   200 ml   1 litre   900 ml

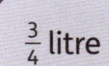

14 Tito has a 1-litre jug that is $\frac{7}{10}$ full of water. He then pours out half a litre. How much is left?

15 Today, the temperature is 22 °C at 8 o'clock in the morning. The temperature goes up by 8 degrees during the day and then drops by 5 degrees in the evening. What is the temperature in the evening?

16 What is the most likely temperature of water that boiled 5 minutes ago?

   20 °C   40 °C   70 °C

17 Draw a spinner with 4 possible outcomes that are all equally likely.

Glossary

< sign – 'less than' sign

< sign – 'greater than' sign

12-hour clock – uses the numbers 1 to 12 for the hours from 1 o'clock in the morning to noon (a.m. times) and repeats the numbers 1 to 12 for the hours from 1 o'clock in the afternoon to midnight (p.m. times)

24-hour clock – uses the numbers 0 to 23 for the hours in a day; 01:00 is 1 o'clock in the morning or 1 a.m.; 13:00 is 1 o'clock in the afternoon or 1 p.m.

A

analogue clock – a clock that shows the time using hands that point to numbers arranged in a circle

anti-clockwise – the opposite direction to clockwise

area – the amount of space taken up by a 2D shape, measured in square units such as cm²

array – objects or pictures arranged in equal rows and columns; arrays can help us multiply

B

bar chart – a chart with bars that show quantities or numbers so we can compare them

C

cardinal points – the four main directions on a compass (north, south, west and east)

Carroll diagram – a table for sorting items according to two sets of categories

centimetre (cm) – each centimetre is made up of 10 millimetres (10 mm)

circle – a round 2D shape

clockwise – the direction the hands move around a clock face; the opposite direction is called anti-clockwise

compasses – a mathematical instrument used for drawing arcs and circles

cone – a 3D shape with a pointed end, a curved surface and one circular face

cube – a 3D shape with six square faces

cuboid – a 3D shape with six rectangular faces

cylinder – a 3D shape with two circular faces and one curved surface

D

day – one of the days of the week, for example, Monday or Thursday; each day lasts 24 hours; there are 7 days in 1 week

decimal point – the point between dollars and cents or pounds and pence; we write four dollars and 25 cents as $4.25

degree (angles) – the unit used for measuring the size of angles; a square corner is a ninety-degree (90°) angle

degree (temperature) – the unit used for measuring temperature; for example, we write 40 degrees Celsius as 40° C

denominator – the number at the bottom of a fraction, which shows how many equal parts the whole is shared into; 3 is the denominator in $\frac{1}{3}$

difference – we find the difference between two amounts by subtracting; the difference between 4 and 6 is 2

digital clock – a clock that shows the time using numbers and dots; the digital time 07:00 is the same as 7 o'clock

divide – to share or group a number into equal parts; for example, 15 ÷ 3 is the same as 15 shared into 3 equal groups

rectangle – a polygon with four sides and four square corners

regroup – rewrite a number using a different place value to make it easier to add or subtract; for example, change a ten to 10 ones

regular partition – to share a number into its place values; for example, $142 = 100 + 40 + 2$

regular polygon – a straight-sided shape with equal side lengths and equal angles

remainder – the amount left over after dividing into equal groups; 5 divided by 2 makes two equal groups of 2 with a remainder of 1

right angle – a 90-degree angle, a square corner or a quarter turn; four right angles make a full turn

Roman numerals – an ancient number system that uses seven letters to represent numbers

round – to change a number to make it easier to work with; for example, we can round a number to the nearest ten

S

second – when a clock ticks, each tick is one second; there are 60 seconds in 1 minute

sequence – an ordered list of numbers, shapes or pictures that follow a rule; also called a pattern

set square – a triangular instrument with a right angle that we use for mathematical drawings

square – a polygon with four square corners and four sides with equal lengths

square corner – an L-shaped angle or corner, such as the corner of a square or rectangle; also called a right angle or 90-degree angle

square units – the units for measuring area; squares that we use to work out how much space a 2D shape takes up

square-based pyramid – a 3D shape that has a square base and four triangular sides that meet at a point

strategies – ways of working something out

sum – the total when we add two or more values together

T

tally mark – a small mark used to count one object; we draw tally marks in groups of five, like this 卌

tally table – a table we use to record data using tally marks to make it easier to count items

temperature – the measure of how hot or cold something is; we measure temperature in degrees Celsius

tenth $\frac{1}{10}$ – when we share one whole equally into 10 parts, each part is one-tenth

tenth (10th) – an ordinal number; the number 10 in position order; for example, the 10th of January is day 10 in January

term – each number, shape or picture in a sequence or pattern

term-to-term rule – the rule for generating the next term in a sequence or pattern

thermometer – an instrument for measuring temperature in degrees Celsius (° C)

third $\frac{1}{3}$ – when we share one whole equally into three parts, each part is one-third

third (3rd) – an ordinal number; the number 3 in position order; for example, the 3rd of January is day 3 in January